Staff-Less Libraries

Chandos

Information Professional Series

Series Editor: Ruth Rikowski
(email: Rikowskigr@aol.com)

Chandos' new series of books is aimed at the busy information professional. They have been specially commissioned to provide the reader with an authoritative view of current thinking. They are designed to provide easy-to-read and (most importantly) practical coverage of topics that are of interest to librarians and other information professionals. If you would like a full listing of current and forthcoming titles, please visit www.chandospublishing.com.

New authors: we are always pleased to receive ideas for new titles; if you would like to write a book for Chandos, please contact Dr Glyn Jones on g.jones.2@elsevier.com or telephone +44 (0) 1865 843000.

Staff-Less Libraries

Innovative Staff Design

Carl Gustav Johannsen
Royal School of Library and Information Science
University of Copenhagen
Denmark

ELSEVIER

CP
CHANDOS
PUBLISHING

An imprint of Elsevier • elsevier.com

Chandos Publishing is an imprint of Elsevier
50 Hampshire Street, 5th Floor, Cambridge, MA 02139, United States
The Boulevard, Langford Lane, Kidlington, OX5 1GB, United Kingdom

Library of Congress Cataloging-in-Publication Data
A catalog record for this book is available from the Library of Congress

British Library Cataloguing-in-Publication Data
A catalogue record for this book is available from the British Library

ISBN: 978-0-08-101923-8 (print)
ISBN: 978-0-08-101924-5 (online)

For information on all Chandos Publishing publications visit our website at https://www.elsevier.com/books-and-journals

Publisher: Glyn Jones
Acquisition Editor: Glyn Jones
Editorial Project Manager: Charlotte Rowley
Production Project Manager: Poulouse Joseph
Designer: Mark Rogers

Contents

List of Figures

List of Tables

List of Boxes

Acknowledgments

This book is very much a product of the stimulating and open intellectual environment of the Royal School of Library and Information Science at the University of Copenhagen where I have been privileged to work for many years. Among my many remarkable colleagues who have contributed, directly or indirectly, with inspiration or comments, I will especially thank Dorte Skot-Hansen, Vice-Director, Hans Dam Christensen, Henrik Jochumsen, Lennart Bjørneborn, Nan Dahlkild, Nanna Kann-Rasmussen, and my Swedish, Ph.D., student, Lisa Engström. The staff of the university library and especially Karen Margrethe Ørnstrup and Lisbeth Rasmussen have also delivered valuable inputs and guidance.

I would also like to thank members of my national network of library professionals and researchers for their valuable contributions: professor Gunnar Lind Haase Svendsen, Hellen Niegaard, Jens Thorhauge, and Jonna Holmgaard Larsen and special thanks for contributions to Carsten Nicolaisen, Sønderborg, Finn Petersen, Copenhagen, Gitte Fisker, Aalborg, Hanne M. Sørensen, Sønderborg, Jon Madsen, Bornholm, Kirsten Boelt, Aalborg, Martin Lundsgaard-Leth, Ikast-Brande, Mogens Larsen, Silkeborg, and Pia Henriette Friis, Kolding.

Besides, a number of professionals and researchers from different countries have contributed: Andreas Vårheim, Norway, Ane Landøy, Norway, Annelie Krell, Sweden, Cecilia Ranemo, Sweden, Claudia Knauer, Germany, Ibi Engsby, Norway, Kristiina Kontainen, Finland, Kristina Elding, Sweden, Leikny Haga Indergaard, Norway, Lisa Roberts, Peterborough, UK, Marit-Gro Berge, Norway, Mikko Vainio, Finland, Sven Arne Tinnesand, Norway, and Svanhild Aabø, Norway.

I would also like to thank leaders and staff from the two, recently merged, library software companies Bibliotheca and Cordura for their invaluable inputs, data, and contributions: Anja Høyer Bæk, Aarhus, Lau Rasmussen, Aarhus, Richard Stewart, UK, and Sven Mønsted Hilm, Copenhagen.

For continuing interest in the project and valuable support, I would like to thank the publisher, Elsevier, and especially George Knott, Poulouse Joseph and Tessa de Roo.

Book projects cannot be completed without sacrifices on the home front. Here I would like to thank the following people for their patience and support: my grandchild, Clara, my son Anders who recently has moved to Cambridge, my daughter Rikke, and my girlfriend Gitte.

Østerbro, Copenhagen, Denmark
October 2016

Introduction

1

Purpose

The purpose of this book is to present and illuminate both the challenges and the benefits associated with a quite new type of public library: the staff-less or the open library or the unstaffed library. The terms "staff-less" and "open" will be used synonymously from now on.

The first open library was opened in Denmark in 2004 and since 2010, the staff-less model has spread in Denmark and to the other Nordic countries and to other countries in Europe, such as the Netherlands and the United Kingdom. At the same time, the model has also emerged in several Far Eastern countries such as China, Japan, Singapore, and Taiwan.

Staff-less libraries are found both among academic and public libraries. However, there are many differences on how they work, their opening hours, and the like; also the contexts associated with the two library types are quite different. Therefore, this book will, primarily, focus on the staff-less public libraries. However, when interesting parallels can be drawn, relevant academic library experiences will be referred to also. For example, because the interest for late-night access to the library seems to be quite different among academic and public library users (Lawrence & Weber, 2012), the causes of such remarkable differences will be dealt with.

Since about 2004 the open library type has evolved in many countries, especially, in the Nordic countries in Europe. Experiences with unstaffed public library services from many countries will be dealt with. However, the book, in particular, will focus on the following seven selected countries:

- Denmark
- Finland
- Norway
- Sweden
- Germany
- United Kingdom
- United States

These countries have, however, not been chosen because they were the only countries where the open library concept has been tested. Asian countries such as China, Japan, Singapore, and Taiwan have already been mentioned as early adopters of the necessary radio frequency identification (RFID) technology in public libraries. Therefore, it has been a combination of languages available for the author and countries with relevant experiences that has determined the selection. For example, the Dutch experiences have been omitted because of language difficulties.

Although many users, local politicians, national library authorities, and library leaders and staff members have welcomed the open libraries, resistance and critique have

Staff-Less Libraries. http://dx.doi.org/10.1016/B978-0-08-101923-8.00001-0

also been submitted by professionals, certain associations, and even by some users. This important aspect is clearly mirrored in the subtitle of this book: *Navigating the Innovative Staffing Design*, where I hope that the facts and reasons presented during the following pages will contribute to constructive dialogs in the libraries and wherever else the open library issue is discussed to facilitate the navigation toward innovative staffing designs.

Definitions

The definition of the term also demands attention, whereas the risk of misunderstandings is at hand. As far as possible, the specific term used in the country in question will also be used in the book. In Sweden and in Norway, for example, the term "more opened" instead of "staff-less" or "open" is frequently used. Also in Denmark, staff-less libraries are more often referred to as "open libraries" than as "staff-less" or "unstaffed" libraries. Such preferences have something to do with the negative connotations associated with terms such as "unstaffed" and "staff-less."

Because, however, all the applied terms "staff-less," "open," "more-opened," and "unstaffed" library, immediately, could appear somewhat misleading, a further, more elaborated, definition seems to be necessary. Below, however, the four terms, will, more or less, be used synonymously. As it will emerge from the next chapter or Chapter 2, a fifth term, "self-service," however, will be avoided or handled with care because it indicates a somewhat different status compared to the terms "open," "more-opened," "staff-less," and "unstaffed." Also, in the business world, for example, most "self-service" super markets are not "staff-less" at all.

What Is an Open Library?

What is an open or more-opened or staff-less or unstaffed library? Is it, for example, the same as a self-service library? It is certainly not the same. But why is it not the same? Because the latter question is more complex than immediately thought, it cannot be answered fully by a straight "yes" or "no." To consider the question about choosing a proper terminology, I will start the discussion by using a small bibliographic exercise to show the character of the existing ambiguities and the complexities of the issue.

Open Libraries in Bibliographic Databases

In online databases, especially the ones prepared by professional librarians, you usually expect to find well-defined and precisely controlled keywords concerning the important concepts. Let us take a look at such a professional database and consider if that is also valid for the conceptual framework around terms such as "staff-less" or "open" libraries.

Library and Information Science Abstracts

If you make an online search in *Library and Information Science Abstract* (*LISA*), one of the main databases of the library profession, you would get 65 hits if your

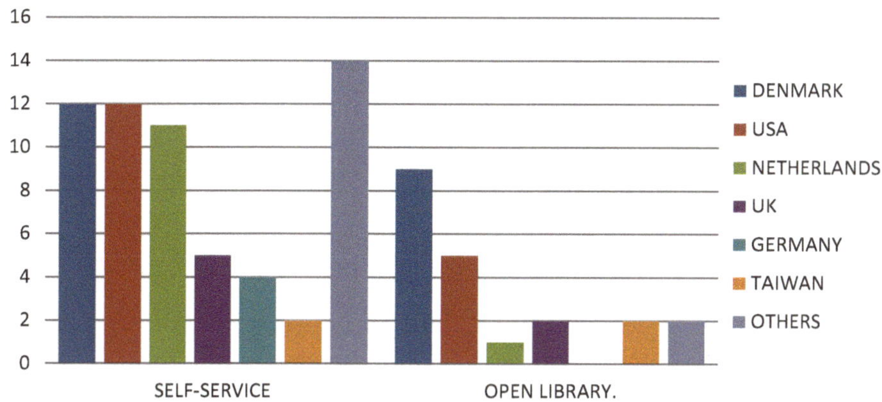

Figure 1.1 Literature on Open Libraries 2016 unit: articles—N = 62. Library and Information Science Abstracts (LISA), August 16, 2016.

search was made on August 16, 2016. My point here, however, is that most of those 65 records, actually, do not deal with the same topic as this book. The database search was made by applying the available two controlled search terms:

- "self service" AND "public libraries."

The controlled term, "self service," has been used by LISA since, at least, the beginning of the 1980s.

Apart from three obviously irrelevant hits, the remaining 62 records deal with public library experiences in 17 different countries in Europe, Asia, and America as shown in Fig. 1.1. The decision whether an article was about self-serviced or about staff-less practices was mostly based on the title or the abstract.

We will now take a closer look at some of the retrieved records. An article such as, for example, Pamela Smith's "*Self-Check: A Lesson in Mistaken Assumptions*" (2008) about self-check machines in libraries in New Hampshire is considered to be nonrelevant because, based on the abstract, it is clear that the libraries in question are staffed. Indeed, automated self-delivery and self-checkout of library materials were already started in the 1990s when clerical functions associated with delivery and loan began to be automated. Many articles from different countries on automated delivery or checkout system where the "self-service" keyword has been applied are mentioned in Table 1.1:

Apparently, articles published before 2004, when the first modern open library appeared, are nonrelevant in the present open library context. The relevance of the post-2004 titles here is determined based on the abstracts.

Borderline Cases

A more problematic open library relevance occurs in texts such as Pamela Karl's article (2011) on the collaboration between Boopsie, Inc., and Cuyahoga County

Table 1.1 LISA Records About Self-Service Libraries

Baruchson-Arbib (1997)	Israel
Hokka-Ahti (2005)	Finland
Jakobsen (1996)	Norway
Kjekstad (1996)	Norway
Kraljic and Maver (2006)	Slovenia
Nyeng (1998)	Denmark
Pohl and Schubert (2006)	Germany
Palmer (2006)	United Kingdom
Petersson (1996)	Sweden
Steffensen (1992)	Australia
Thompson (2006)	United States
Van de (2002)	The Netherlands

Public Library (CCPL) in Ohio. In the article, it is demonstrated how open dialog and an exchange of ideas can turn into win–win solutions for libraries and vendors. It describes the CCPL Mobile smartphone app and the Boopsie Book Check feature that allow patrons to check out materials using their smartphones. Here, the self-service dimension is clear but it is less clear whether the solution also works outside the staffed opening hours. Because of this uncertainty, this and similar other articles, therefore, are also classified as nonrelevant, as open library texts.

Another borderline case is, for example, about Michele Hampshire and Cathy Sanford's article, "*Library-a-Go-Go: Bringing the Library to the People*" (2009). This article is about the use of ATM-style technology adapted to lend books in the San Francisco Bay Area. Similar examples are found in Sweden and Germany. Here, it is another important element of the open library, namely, the physical library, that is missing, whereas the self-service aspect is clearly at hand here. Indeed, there is, certainly, no staff posted to help users operate the book vending machines outside the library. Therefore, the "Library-a-Go-Go" article title is considered relevant even if the interior of the library is not used.

Self-Serviced Book Mobiles

We have already argued that a book vending machine could be included in the category "open libraries," although it is evident that such a mechanism is not a real library. But then what about staff-less—but, hopefully, not driver-less—book mobiles? A 2006 article, "*Mountain View Implements Innovative's Self-Service Products*" deals with a self-service product that enables the mobile library (book mobile) to allow users to check out books and register for library cards on their own through real-time access to the Millennium integrated library technology platform. The system works by equipping the mobile library vehicle with two laptops outside the vehicle where users have wireless access to Millennium and the Internet. Indeed, a book mobile seems to contain more library spirit than an isolated book vending machine. Therefore, this and similar texts are also included.

The result of applying the algorithms described earlier shows that only one-third (about 20 out of 62 relevant titles) actually deals with open or staff-less libraries in the sense used in this book. Even among the 20, there are gray area titles, as the two mentioned earlier. Many of the screened titles, especially, titles from before 2004 deal with different applications of self-checkout systems in different countries.

Same Pattern Found in Library and Information Technology Abstracts

Library and Information Technology Abstracts (LISTA) is another important library and information science database, focusing more on information technology (IT) than LISA. A similar search as the one in LISA was conducted on September 6, 2016, in LISTA:

Public libraries (subject term) AND self-service

The search produced 58 hits. Among them a clear majority dealt with libraries with automated loan and checkout systems installed, although the choice of words could confuse the reader. "*Great Britain's first totally self-service library*" opened at the London Borough of Sutton in 2004 and it was definitely not a staff-less library, although the word "totally" indicated something very special (Anonymous, 2004). In other sources, expressions such as "completely self-service Twenty-four Seven" in the new city library of Newcastle Upon Tyne in the United Kingdom refers to a real staff-less library because it is mentioned that it is opened "even during closing times" (Venuda, 2009). Among the many articles from different countries on automated delivery or checkout system or self-service kiosks, where the "self-service" keyword has been applied, I will, here, only mention spots—in chronological order—that have not already been listed among the LISA records (Table 1.2):

All of or nearly all of these articles—all published after 2004—deal with implemented self-service system in public libraries and not with open or staff-less libraries.

Table 1.2 LISTA Records About Self-Service Libraries

Rogers (2005)	Woodstock Public Library (USA)
Riess-Taggart (2006)	Palm Beach (Florida, USA)
Anonymous (2008)	Paris (France)
Anonymous (2009a)	San Diego (California, USA)
Anonymous (2009b)	San Francisco (California, USA)
Anonymous (2009c)	Sunnyvale California public library (USA)
Anonymous (2009d)	Affoltem regional library (Switzerland)
Anonymous (2011a)	August Cesare library, Zagreb (Croatia)
Hill (2010)	Jacksonville public library, Florida (USA)
Anonymous (2012b)	Ottawa public library (Canada)
Enis (2014)	Queens library, New York (USA)
Sigwald (2016)	Baltimore county library (USA)

The Importance of the Distinction Between Open and Self-Service

But why is this distinction between self-service and staff-less important at all? Is it a difference that makes a difference? It is important because only the staff-less or open library technologies in practice allow the library to offer its users prolonged weekly opening hours in the order of 60–80 or more hours per week, which the self-serviced library seldom does. Indeed, self-service technologies where the staff is still present in the library do not, usually, allow that many opening hours.

Self-Service and Open Libraries Are Different in Several Respects

Besides the prolonged opening hours, the open or staff-less library model also challenges the library in a much more radical way than the self-service library model does. While the staff is still present in the self-serviced library, the staff-less reality is much more demanding as to the responsibility and the nondestructive behavior of the users during the unstaffed opening hours. The circumstance that a self-service library has functioned well in a given local community is, therefore, not at all a guarantee that a staff-less library would do the same.

The two lists of countries where self-serviced public libraries have been implemented derived from database searches in LISA and LISTA contain both more countries and, in particular, more libraries than a list of countries with implemented staff-less libraries. This eye-catching difference suggests that self-service and staff-less libraries lead to quite different challenges.

The aspects involved are, for example, also quite different. The users feeling safe and secure in the library are usually not an urgent need in most libraries during the staffed opening hours. In the open library, it is a challenge of primary importance. Also values such as "social capital," "ownership," and "trust" could certainly be useful but they are not absolutely essential in a self-service library. Like commercial supermarkets, self-service libraries can be installed nearly everywhere. In staff-less libraries, the values mentioned earlier are more or less essential because the users here are totally left to fend for themselves.

But Self-Service Technologies Are a Necessary Prerequisite

Self-service technologies are, nevertheless, a necessary prerequisite for today's open library revolution. Thorhauge (2011) has emphasized that a good reason for the success of the open library model in Denmark is that:

> *for many years Danish libraries have worked with self-service in loan and return transactions which means that a majority of users are familiar with self-service procedures which by the way are extremely simple.*

A similar observation was done in the Peterborough, UK, by Lisa Roberts, Strategic Client Manager: Culture and Leisure at Peterborough City Council:

> *With so many of our patrons already accustomed to using self-service, accessing the library by simply scanning their library card and entering their usual pin*

number at the external access control panel was easy enough for them to adapt to straight away

Bibliotheca open+ brochure (2016).

As extra systems are needed for an open library, it can be mentioned that the access systems to enter the library by the use of a card are found in nearly all open libraries. Closed-circuit television (CCTV) and video surveillance are also considered to be essential accessories, although most open libraries in Sweden seem to do well without video surveillance. Consequently, the term "self-service" will not—unlike terms such as "open," "more opened," "staff-less," or "unmanned" and the like—be used synonymously when dealing with open or staff-less libraries. Thus the terms open or staff-less or unstaffed library contain two dimensions:

- Open-ness—typically, prolonged opening hours
- Staff-lessness—staff-services not available

"Open" or "More Opened" Are the Preferred Terms

The term "open" or "more-opened" library is the preferred term in Denmark. It is also the preferred term in this book, although alternative terms such as "unstaffed" and "staff-less" are also used to create some linguistic diversity. Self-service, on the other hand, will not be used except in cases where the context is specified as self-service and not as staff-less.

In Denmark, we also find some of the earliest, if not the earliest example, of practical implementations of the concept. Jens Thorhauge, the former director of the Danish government for Libraries Agency and Media, explains why the term "open" was chosen.

In the first instance the term 'open library' was chosen as the concept was implemented in libraries that had typically been branches with short—and often inconvenient—opening hours. And 'open library' is also preferable to 'self-service library', as this concept requires another kind of professional support enabling the user to complete the necessary transactions themselves and to find their way in the library.

Thorhauge (2011).

Because the open library concept, especially, in the beginning, was mostly implemented in libraries that had typically been branches with short and often restricted and inconvenient opening hours the word "open" was likely to signal more openness in terms of more opening hours. In Finland, Norway, and Sweden the term "more-opened" is usually used emphasizing that the staff-less opening hours only make up a part of the total opening hours of the library.

Open Often Means (Nearly) Always Open

The extended opening hours were seen as an important step to create a more open library understood as a library with an increased availability. Maybe to further clarify the term "open" one could add the word "always"—the "always open library"—to indicate that 24/7, twenty-four-seven, opening hours (24 hours day and night—7 days per week) was both the ambition and the ultimate goal. Most public libraries, however, usually, would only practice opening hours from early morning to late evening and, typically, not during the night. Some academic libraries and, especially, their patrons, on the other hand, have placed high value on late-night access to the library (Lawrence & Weber, 2012). A variant of the "open library" term is the "more opened" library, which is used in the other Nordic countries. This term signals that the library is more opened than usual. A staff-less library does not, necessarily, have more opening hours than a staffed library, although it often has.

Open Does Not Mean Unlocked

Furthermore, "open" does not indicate that the library doors are open, that is (to say), left unlocked, so that everyone can enter the facilities during the staff-less opening hours. On the contrary, a card—typically a medical card or another card or specific library card, which all adults possess or can get—is required to enter. Moreover, a password is typically required too. That means, for example, that children without such cards, cannot lock themselves in, they can only enter the library in company with adults. However, there also exists examples of definite unlocked open libraries, for example, in Norway.

Services in the Open Library

When having first entered the library, the visitor can, then, typically, enjoy the full package of public library services:

- deliver already loaned library materials,
- connect to Internet and library catalogs through the library's computers,
- order, pick up, select, and loan books and other library materials, and
- read available newspapers and journals.

Sometimes the available services can be restricted. In Peterborough in the United Kingdom, for example, users cannot take out compact discs and digital versatile discs (Hitchcock, 2016).

Besides, many staff-less libraries offer facilities for meetings, workshops, and the like organized by the library visitors themselves. A lot of libraries also offer—both during the staffed and unstaffed opening hours—opportunities for accompanied children to play, utilize computer games, and perform similar activities. Therefore, the rise of the staff-less library movement is considered by many professionals, local politicians, and not at least by many users as a significant progress and as an improvement of public library services in general. We will later consider why some professional librarians and others, nevertheless, have uttered a critical position toward the staff-less library concept.

Never Staffed Open Libraries Are Rare

As mentioned, terms such as "staff-less," or "unstaffed" libraries, so to say, to a certain degree invite misunderstandings and sometimes resistance too. Also, here some further clarifications are needed. The first obvious misunderstanding could be that a staff-less library should be libraries that are staff-less during all opening hours. The American, Wallin (2015), therefore, emphasizes that unstaffed or staff-less does not mean "never staffed."

Certainly, examples of such totally unstaffed libraries appear; but they are unusual and rare. In Denmark in 2011, there were 81 staff-less libraries; only six (7.5%) of them were totally unstaffed, all year round. Today, there are according to the national library authority 297 open libraries of which only five (1.7%) are staff-less all the time (Slots-og kulturstyrelsen—Oversigt over åbne biblioteker per 1 september 2016/slks.dk).

Typically, a staff-less library is only staff-less during parts of the opening hours. In Denmark, where the staff-less libraries first became a common public library offering, the typical model included a week with 20% staffed and 80% unstaffed opening hours. Since 2011, the percentage of unstaffed opening hours in Denmark has been reduced to 61% (in 2014), whereas the total number of opening hours from 2010 to 2014 has been doubled (Slots-og kulturstyrelsen, slks.dk).

The unstaffed opening hours typically mean that the library is open in principle 24/7, but in reality it is rather 14/7—all 7 days of the week from early morning, for example, 8 a.m. until late evening, for example, 10 p.m. The Swedes, Elding, and Krell (2015) recommend a distinction between "more opened" (both staffed and un-staffed opening hours) and "unstaffed" with only staff-less opening hours. Using their distinctions, 1.7% of the Danish staff-less public libraries were "unstaffed" in 2016.

The Staff Has Important and Necessary Functions in Staff-Less Libraries

Also, the lack of clarity concerning the important role of the staff in staff-less library contexts, contains risks of misunderstandings. Although the staff members are normally not present in the library during the staff-less opening hours, they perform a number of necessary and vital functions. Besides, ensuring that the library's personal computers (PCs), printers, and other technical equipment work and that the library materials are organized in a user-friendly and understandable way, there are many other ways for the staff to make the library facilities appear inspiring and easy to overview. An example could be through relevant and informative exhibitions of books and other library material and through different kinds of cozy arrangements. Also through new and innovative forms of communication around, for example, literature circles, the professional staff can support user activities during the staff-less opening hours. Thus the roles of the library staff concerning the open libraries are in some respects similar to the tasks of curators in museums. Moreover, for example, around Christmas, many Danish open libraries, are glad to offer coffee, tea, and biscuits to their users. The role of the professional and clerical staff here also contains host-like features. Indeed, the responsibilities of the staff cover much more than just cleaning up before and after the staff-less hours.

An Operational Definition of an Open or Staff-Less Public Library

To sum up we will conclude that despite a considerable variability, an open library will typically possess the following key characteristics:

- Gives access to the library through the different types of identity cards through an electronic access system.
- All the services of the library are available during the unstaffed opening hours (loan, delivery, payments, browsing, reading, meeting place, and the like).
- The opening hours are usually significantly prolonged approaching the 24/7 ideal.
- The physical library is typically—but not at all always—for example, in Sweden, only in exceptional cases—video monitored through installed cameras.
- Staff members can be present in the library during the staff-less opening hours—they are, however, not committed to offer services to the users.
- Volunteers might also be present in the library during the unstaffed opening hours (the presence of the volunteers, however, does not mean that the library is considered to be staffed).

We will now take a look at some of the problems or challenges that face the open libraries.

Challenges

Certainly, staff-less library services as well as staffed services contain a number of challenges. To ensure the security and safety of the visitors during the unstaffed opening hours belong to the most important. Here, it makes sense not only to consider the objective but also the subjective dimensions—the feeling of being safe. A number of means serve to achieve this purpose. They will be dealt with, in further detail, in Chapter 5. Security and safety, for example, could be achieved through electronic surveillance devises, placed at strategic places in the library. However, also, the non-electronic aspects as regards, for example, the architectural design of the library are relevant to be aware of. Eliminating the dark or messy corners, for example, could certainly help to ensure the peace of mind of the visitors as well as to improve the user friendliness of the library. Besides, removing or reducing different risks, libraries could also do a lot to make the facilities more inviting through, for example, comfortable furniture and appropriate working places for typical library visitor segments. Students, thus typically, demand other kinds of facilities such as appropriate working tables than elderly, leisure-time visitors. In any case, such installations are not free and should be carefully planned.

Indeed, many of the tasks associated with furnishing and designing the library for the staff-less opening hours are very much like the similar tasks for the staffed hours. However, the staff-less context tightens up demands for a both logic and intuitively way of organizing the library's materials to make searching for specific titles easy and convenient without the need for personal assistance. At the same time, demands for making the library a place where the users will get inspiration and new ideas also represent an important challenge in the staff-less library.

Serendipity in the Open Library

The library should be able to meet the diverse challenges of both fast and predictable access to the materials and the demand for a library design increasing serendipity. Serendipity refers to the user's chance for making—at the same time—useful and unexpected discoveries in the library. Normally, library services are aimed at answering specific questions, whereas serendipity is much more about making unexpected findings—something you didn't know that you wanted. Principles and examples of how to achieve such purposes will be further dealt with in Chapter 6.

Nudging

Nudging (Economist, 2014; "Nudging" Wikipedia, 2016) is an interesting approach that has been recommended for open libraries too. Yet because nudges do not impede personal choice entirely and are considered to be "freedom-preserving" and empowering (Hall-Ellis, 2015; Thaler & Sunstein, 2009, p. 255), individuals maintain their option also to make poor decisions and choices. Nudge theory is based on ethical behavior and encourages free choices for each individual that can result in better thinking, behavior, and decisions. The arrangement of the physical space and messages to the users can be utilized as nudges. The small fly often seen in the bottom of urinals is an often used example of an effective nudge. There are many different nudges as, for example, temptation and spotlighting, which can be used in open libraries. The widespread exposure of book title pages on slanted shelves in open libraries can also be used as an example. The case study in Box 1.1 contains a practical example of the use of nudging in staff-less public libraries.

We will now take a look at some of the benefits created by the new open library concept.

Benefits

In short, the open libraries represent an effort to adapt library services to the fast changing requirements of a modern society by utilizing recent technological opportunities. A staff-less library is, therefore, much more than simply a traditional library with prolonged opening hours. Rather the concept represents an intelligent utilization of modern Information and Communication Technology (ICT) and in most cases also CCTV surveillance technology to enable public libraries to deliver an individualized and customized on demand service to its users. As we will see, the justifications for introducing open libraries in a specific municipality or in a particular library unit can vary a lot from a step to improve the library's services to pure budget cut purposes.

Similarities With Film Streaming Services

Within a related field, television and film streaming services, we have observed a similar customizing trend, where the opportunity to watch and enjoy ones favorite TV program or film is no longer restricted to specific hours of the day. In contrast, the films

Box 1.1 Staff-Less Libraries on the Island, Bornholm, Denmark: A Case

A couple of years ago I told a Polish library director about our local strategy of converting all public libraries on the island of Bornholm into staff-less libraries: access to the library all day with partly staffed hours and partly unstaffed self-service. He shook his head at the idea and told me it would be an enormous risk, and that this would never work in Poland. However, the introduction of this concept proved the biggest success of the island's seven public libraries for a number of reasons:

1. we put our trust in the hands of the library users, which they appreciated and lived up to;
2. they were able to help themselves all day and go through formerly closed doors on weekends, which many families had wanted to do for years; and finally
3. we maintained the manned hours of staff so the users would have a choice between service and self-service.

The results could be seen immediately in the libraries' statistic sheets: in the first year the number of visitors rose by 20% and the next year saw an increase in the number of lendings of 10%. News spread quickly around the island, which meant that our satisfied customers would act as ambassadors and attract new customers. People generally regarded the possibility of going to the library at any time of day as an enormously increased level of service.

Our success was based on experiences from other municipalities in Denmark, stating that the risk of anything going wrong would be minimal in a small community. So we created a success by starting out with a conversion of the smallest branch libraries on the island. When these people got the new system first, they felt privileged, and this created a demand for the system among the users in the bigger communities. After 2 years, only the two biggest libraries on the island remained, as we would have to make a few alterations in the building to make sure that the users would feel comfortable and safe in a big library with no staff. After another year, all libraries ran smoothly in their new status.

We even carried out a few experiments regarding the impact of nudging in a staff-less library: to minimize flaws at the self-service automats small signs encouraged people to seek each other's assistance. This immediately reduced the number of alarms by one-third.

It seems that we created a win–win situation between staff (no lost jobs), users (extended access to the library), and local politicians (appreciating extra performance at no extra cost). As people gradually got used to self-service, the lending activities would also extend to the weekends, the activities in staff-less periods outnumbering loans within the manned hours in the smaller branches. In this way, we were able to gradually change priorities from direct customer service to cultural arrangements and background work of getting the materials ready for the customers.

Box 1.1 Staff-Less Libraries on the Island, Bornholm, Denmark: A Case—cont'd

The main point I want to make is that showing the customers that we trust them is the key to success. It is also a far better investment than fearing that a small minority might ruin it all by means of havoc and plunder, which my Polish colleague had considered inevitable. We have had had no incidents that could not be handled easily, due to records of people who entered the library, discreet camera supervision, and an advanced gate system that will keep a record of materials leaving the library without registration. But of course, this recipe cannot just be copied everywhere. Bornholm is generally a cozy place to be, closer to the end of the world than it is to a capital with huge conflicts and social problems that might lead to a different behavior and might demand different solutions.

By Library director **Jon Madsen**, Bornholm Public libraries, Denmark.

or broadcasts have become accessible on demand whenever the user wishes to watch it. In the same way, the opportunity to visit a physical library on demand, from early morning to late in the evening, has become a realistic option in many cities and local communities. Compared to earlier times about just 10–12 years ago, the moments where a citizen could choose to visit and utilize the opportunities of a professional library system have been heavily enlarged. This is why many users and librarians consider the staff-less library as a step forward as to user friendliness. Another point of similarity between streaming services and open libraries is that both the TV on demand and the open library are made possible by a combination of new technological opportunities and market demands.

A similar combination of both modern ICT and traditional physical elements is found in many different contexts, versions, and applications in the modern society. Buying on the net, for example, also contains a combination of both virtual and physical components, widening the opportunities of the customers. However, as will be argued in the next chapter, the staff-less library concept is also rather unique in a number of respects. We will now consider the benefits at three different levels: the individual, the organizational, and the societal level.

Library Service on Demand—The Individual Level

At the first level, the individual, the "open library" concept represents a clear step toward a more individualized library service delivery system. Indeed, the on demand character of library visits clearly indicates an increased individualization. A user who is in the mood to use the library a late Sunday evening can do so now, without being restricted by traditional opening hours. She can, indeed, choose to get access to knowledge and information in books and printed materials when she needs it. Or she can make use of the opportunity to just relax in the often inspiring surroundings of the local public library. Although online access to

Box 1.2 Purposes of Offering Open Library—Gladsaxe Libraries (Denmark)—Case

Esben Fjord, Head of Development at Gladsaxe Libraries, comments: "We wanted to give our citizens the opportunity to visit the library when they wanted to; there are a lot of people that live in this community that are at work all day, so when the library closes at 4 p.m. they are unable to visit the library until the weekend—this was something that we wanted to change
Esben expands, stating: "Flexibility is really where we have seen the biggest improvement—not just for us, but for our patrons. People can now come and visit the library whenever they want, whether this be to check out or check in a book, to keep in touch with people using the PCs, to have group or community meetings or just to read the paper—the open+ concept has worked really well at keeping us relevant in the community.

Open+ brochure—issued by Bibliotheca (2016).

library materials predates the open libraries, the access to printed materials and the opportunity to stay surrounded by the inviting public library premises outside the staffed opening hours represent a significant library service step forward. It should be emphasized that open libraries are not only about spontaneous and transient wishes. Many commuters, for example, are unable to use a library during its normal opening hours. Box 1.2 deals with flexibility as a core benefit created by the open library model.

Meeting Places—The Organizational Level

The open libraries, however, also offer opportunities for the individual to socialize with others. The way you socialize in a staff-less library context can, obviously, be less intensive than practices in other urban or rural meeting places, such as pubs and sports clubs. Later we will consider in more detail, how staff-less libraries by some researchers are seen as valuable so-called "low-intensive" (Audunson, 2005) meeting places, possessing the specific qualities and societal roles of such fora.

Supporting Local Networks—The Organizational Level

We have now considered the opportunities created by open libraries on a mainly individual level. On an organizational level, staff-less libraries can support local networks and associations by offering meeting room facilities and communication equipment. Often such facilities are separated from loan and reading rooms to avoid disturbing other contemplative and noncommunicating library visitors. Although, the same opportunities also are at hand during the staffed opening hours, the highly extended unstaffed opening hours add a significant degree of flexibility and ability to meet user demands for meeting places. It is emphasized by supporters of staff-less library that both the legitimacy and the use of the public library in local communities are increased as a consequence of open libraries (Elding & Krell, 2015). The legitimacy gained is to a large extent also related to the democratic role of the public library.

Social Capital Creation—The Societal Level

Also at a higher societal level, open libraries contribute. Creation of "social capital," in the tradition of the American sociologist Robert D. Putnam's (1941–) sense of the word, is identified as an important function of both staffed libraries and unstaffed libraries, as noncommercial "third places" (Putnam, 1995). Putnam has suggested the following operational definition of social capital as

> *an accumulated stock of networks, norms and trust, that facilitate co-ordination and co-operation for mutual benefit*
>
> *Haase Svendsen (2013, p. 53).*

Another version of Putnam's social capital definition sounds:

> *By "social capital", I mean features of social life—networks, norms, and trust—that enable participants to act together more effectively to pursue shared objectives*
>
> *Putnam (1996, p. 34) and Miller (2014).*

Many studies that have focused on the relationship between public libraries and social capital (Aabø & Audunson, 2012; Haase Svendsen, 2013; Vårheim, 2007) claim that public libraries are "eminent providers" of human and social capital in small rural communities, where local noncommercial meeting places have become increasingly rare in many countries. It is, furthermore, widely believed by researchers within the field that both the staffed and the open libraries contribute to the creation of social capital and trust. Moreover, empirical surveys have tested and found that public libraries are the breeding grounds for different types of social capital (Haase Svendsen, 2013, p. 52).

Do Open Libraries Also Contribute to the Creation of Social Capital?

It is obvious that a closed down branch library will no longer be able to contribute to the creation of social capital; but what about a branch library where the staffed hours has been reduced from, for example, 40 to 23 h per week and the unstaffed hours increased from 0 to 80 h, spread over all 7 days of the week? Is this library unit also likely to contribute to local community social capital?

Here, an empirical questionnaire survey reporting on social capital creation at branch libraries in 62 rural communities and prepared in 2009 (Haase Svendsen, 2013) only concluded that the respondents, municipal library managers, clearly confirm that local library branches contribute to the creation of different types of social capital. Therefore, such libraries are of great socioeconomic value.

This and similar studies, however, tell little about whether such beneficial effects also are likely to occur in a situation with many staff-less opening hours? This unclarified situation, is probably mostly a consequence of the original purpose and background of the earlier surveys. Indeed, the original purpose was, naturally, not to investigate the effects of open libraries, but rather to examine the consequences of large-scale closures of small branch libraries in rural areas. Indeed, from 1988 to 2011, the number of Danish public library units was reduced from 788 to 352, which represents a 55% decline. At that time, however, the staff-less library evolution or revolution had hardly started. Therefore, the staff-less library aspect only played a minor part in the 2009 survey. A similar picture of closures of small library units took place in many other countries than Denmark.

Hypotheses About the Relationship Between Open Libraries and Social Capital

However, no significant research results yet points at the particular relationships between open libraries and social capital. One hypothesis could be that social capital development is highly dependent upon the presence and activities of qualified staff and volunteers. According to the logic behind this hypothesis, social capital is not likely to flourish during the unstaffed opening hours in a library deprived of the vital resources provided by staff and volunteers.

Researcher Opinions

Because there was no evidence-based research to qualify and test the hypothesis mentioned, I decided, instead, to e-mail interview three leading researchers within the library and social capital field. The chosen researchers were

- The Norwegian professors, Andreas Vårheim and Svanhild Aabø
- The Danish sociology professor, Gunnar Lind Haase Svendsen

All three have published internationally concerning the topic (Aabø & Audunson 2012; Haase Svendsen, 2006, 2013; Vårheim, 2007, 2009, 2011; Vårheim, Steinmo, & Eisaku, 2008). The question they were asked by me was

> *to what extent are libraries—during the staff-less opening hours—likely to contribute to the production of social capital?*

The Unanimous Researcher Responses

The answers from the three researchers were quite unanimous. First, none of them claimed that the evolution of social capital was reserved for staffed libraries. On the contrary, Haase Svendsen emphasized that staff-less libraries could be more productive than staffed as to social capital, if they were used by locals for meetings and events. Also Vårheim's reasoning pointed out that the extended opening hours would rather contribute to social capital creation because the number of users would increase and because the staff-less context would work as an incentive to information sharing and co-operation. Finally, Aabø paid attention to the function of a certain kind of social capital, bridging social capital, which appears primarily in so-called low-intensive meeting places. Low-intensive meeting places are, compared to the high-intensive meeting places, where we can live out our major interests together with people who share them, arenas where we meet people with quite different interests and values (Audunson, 2005, p. 430). Thus it seems as if authorities within social capital creation and libraries agree that staff-less libraries are also able to contribute in a positive way to social capital creation.

We will know take a closer look at some of the underlying sociological mechanisms behind social capital to understand not only that staff-less libraries do contribute to social capital creation but also why they, according to some experts, do so. A recent, interesting study about social capital in public libraries may throw further light on this question.

Social Capital in Copenhagen and Edinburgh

In a comparison between two libraries in two different countries, Scotland and Denmark, focusing on social capital creation in the two cities, Edinburgh and Copenhagen, Miller (2014) has identified three main factors affecting a library's potential to create social capital, namely

1. its buildings and space,
2. its staff and volunteers, and
3. the links that the library has with the community.

Trust Is Decisive

As the article is relatively new, it also contains more substantial reflections about the open library type, which, however, at the time, only occurred in one of the cities, Copenhagen, and not in the other, Edinburgh. The article, however, contains no discussion about whether Copenhagen's staff-less opening hours are likely to increase or decrease the production of social capital. Miller is mostly interested in considering why staff-less libraries are possible at all in Copenhagen; she is, therefore, not so much interested in whether staff-less libraries were likely not only to be based upon but also to produce social capital.

The World Value Survey

Miller's explanation points at the different levels of trust in the two cities as the decisive factor. Copenhagen can offer open libraries because international surveys have documented that Danes possess the world's highest level of trust. Trust is considered to be a key element in social capital. In the World Value Survey (WVS), respondents in more than 100 countries were asked the question:

> *Generally speaking, would you say that most people can be trusted or that you need to be very careful in dealing with people?*

The possible answers stretched from: "Most people can be trusted" to "You can't be too careful" (https://ourworldindata.org/trust).

High Trust and Low Trust Countries

The WVS is interesting in a broader open library social capital context too. First, it is interesting to notice that the countries with the highest level of trust nearly all belong to either the category "historically protestant" (Denmark, Finland, Germany, New Zealand, Norway, Sweden, the United States, and United Kingdom) or to the Confucian (e.g., China, Japan, Taiwan) countries. Ireland is the only catholic country among the high-trust societies. The countries with low scores on trust all belong to either historically catholic or Islamic countries. A high-trust country is, for example, Denmark where 78 out of 100 would agree that "Most people can be trusted," whereas Brazil together with Turkey and the Philippines occupy the bottom positions. Only 3 out of 100 in Brazil agrees that most people can trusted.

A Connection Between High Trust and Open Library Occurrence?

The pattern revealed by the WVS is also interesting because nearly all countries where the open library concept has been implemented—and reported in the library literature—belong either to the protestant category (Denmark, Finland, Germany, Norway, Sweden, the United Kingdom, and the United States) or to the Confucian group (e.g., China, Japan, Taiwan). Within the protestant category, however, considerable variations as to the spread of open libraries can be seen.

The Role of Economy and GNP?

Here, it should also be mentioned that gross national product (GNP) per capita plays a role, but, indeed, a less unambiguous one. Catholic countries with high GNP scores such as Austria, Belgium, and France, for example, score lower on trust compared to similar rich protestant countries. Germany, where the reception of the staff-less library concept has been both late and reluctant because of fear of vandalism, gets a significantly lower WVS score on the trust than the Scandinavian countries. We will later try to explain why the opinions among professionals have been so reluctant; presently (September 2016) there is only one operating staff-less library—around Hamburg—in Germany.

Trust and the Absence of Vandalism

The observation that staff-less libraries imply a high level of trust in the population, has also played a significant role when explaining why the application of the staff-less concept has been so successful in Denmark, especially, in one certain respect: the absence of vandalism. Here, the expectations of many professional librarians were, to begin with, quite pessimistic, anticipating destructive user behavior to be much more common than it showed up to be (Johannsen, 2012a).

The former director general of the governmental Danish Agency for Libraries and Media, Jens Thorhauge, has emphasized:

> *...the concept [the new open library concept] has spread from small villages with a high social control to more complex neighborhoods in cities. And extremely few examples of theft and vandalism such as tagging and aggressive ruining of furniture have been reported. On the contrary, there seems to be a tacit agreement among citizens that this is a good spot for everybody, hard to be against, resulting in a common protection of the library. It seems that the public library has a position comparable to the church that for centuries has been run with open unguarded houses.*
>
> *Thorhauge (2011).*

Although Thorhauge does not explicitly use the terms "social capital" and "trust," his mentioning of concepts such as "social control," "tacit agreement," and "common protection" clearly refers to a framework very similar to social capital. Such an interpretation also indicates that there is a positive relationship between trust, social capital, and open libraries. Putnam has suggested a number of activities to promote social capital creation. These activities include the design of public space to promote casual meetings, and hosting cultural activities to encourage communication between diverse community groups" (Miller, 2014, p. 318). All these suggestions can also be implemented in an open library context.

The Importance of Trust in Libraries

Trust is not only considered a morally valuable concept. Also from an economic point of view, trust makes a difference because costly expenditures concerning contracts and expensive lawyers can be reduced. However, in a staff-less library context trust can also become a very delicate and complex quality, sometimes difficult to manage and handle, which the following example will illustrate.

Generalized Reciprocity

In Miller's comparative study of the two public libraries in Copenhagen and Edinburgh, we find some interesting observations concerning the trust concept:

> *Placing trust in citizens exemplifies Putnam's "generalized reciprocity"; the public must act in one another's best interest, without staff mediation. Discussing group*

> *use of the open library, a member of Ørestad staff stressed the importance of mutual trust: … we don't really have a contract—normally you would perhaps have to sign something so that we know that they're responsible… The Vesterbro manager pointed out that customers expect to be trusted by the library in Copenhagen—some even felt offended by the CCTV in the library, seeing it as a sign of mistrust.*
>
> *Miller (2014, p. 323).*

Is Closed-Circuit Television Surveillance a Sign of Missing Trust?

Especially, the remarks by the Vesterbro manager that the users expect to be trusted and they may feel offended by CCTV equipment in the library, indicate that different kinds of Maslow oriented human needs, apparently, are at play in an open library context. Much focus has been on security, but the Vesterbro experiences may also show the importance of some higher order needs, such as self-realization and identity building (Johannsen, 2012b, p. 341). Similar experiences with users are known from Sweden where a user responds:

> *Staff-less libraries make it possible for the local inhabitants to show how much they really care about their library. It's simply a place to grow*
>
> *Elding and Krell (2015).*

It should be mentioned here that just in Sweden the laws concerning video monitoring in the public space are more restrictive, which means that most Swedish more-opened libraries do not have surveillance equipment installed in the libraries.

In the United States, Wallin (2015) points out, that unstaffed libraries "require mutual trust to succeed."

The focus on trust has furthermore lead to notions of giving the libraries back to the users and to the community as a symptom of lack of confidence in institutions and to replace rules and regulations with trust and a positive mindset (Box 1.3).

Box 1.3 Giving Libraries Back to the Users—Hareskov Library (Denmark)—Case

Anita Bonderup, Head of Department, Hareskov library: when we first introduced the concept and left the library to the patrons, however, there have been no problems at all with anybody mistreating the library when we are not there. In essence we are giving the library back to the community, and they have been taking really good care of it. In fact, the only time we actually had an issue with vandalism or anything like that was when the library was open and fully staffed.

Open+ brochure—issued by Bibliotheca.

Until now, mostly the positive and beneficial qualities of staff-less libraries to the individual user, to the local organizations, and to the local communities have been mentioned. But how has the library profession reacted to the new concept, which might appear, potentially, to make the profession less indispensable. We will take a closer look at this important issue now.

The Pros and Cons

Before considering the arguments in favor of and against staff-less libraries, it is interesting to consider the actors participating in the debate. Why stakeholders are important is because attitudes toward open libraries often are dependent on which stakeholder role you usually play. The principal participants and public library stakeholders are

- Users
- Local politicians
- Newspapers, net media and the like
- Government agencies
- Interest organizations
- Trade unions
- Professional and clerical staff of the library
- Library leaders
- Library researchers

The feelings toward and opinions about open or staff-less libraries are, obviously not, distributed randomly on the stakeholder groups just mentioned. Rather, there seems to be certain recurrent patterns. In general, users, local politicians, government agencies, and some interest associations have been mostly positive toward the idea of staff-less libraries others, especially, among the library staff and some library associations and trade unions, critical voices have been raised in public either against the entire concept of staff-less libraries or against how it has been implemented in practice.

Critical Voices Found in All Countries

The critical voices have appeared in a number of different countries where the concept has been implemented, for example, in Denmark, Finland, Germany, Ireland, Norway, Sweden, Taiwan (Tseng & Kuo, 2009, p. 413), the United Kingdom, and the United States. In the later section on the role of stakeholder the different national variations of the critique and the defense of the open library model will be dealt with in more detail.

The objections, however, have been very different. In some countries, for example, in Germany. professionals have expressed fear of vandalism as their main concern, whereas Swedish, British, and Irish professionals, for example, have been more concerned about the feared erosion of essential job functions and responsibilities of librarians, such as, for example, guidance on the choice of literature and dialog with users on current literal and cultural issues. However, on the other hand, many library directors and librarians have defended and praised the staff-less concept and how it is practiced.

We will now take a closer look upon the central critical positions. Another characteristic trait about the debate on open libraries is that it is often based not on facts but on assumptions and speculations about, for example, what the users "really" need or prefer.

The Main Rebuttals

Four generic types of critique of the open libraries have been identified:

1. the unemployment of the staff counter argument,
2. the economic, budget cut counter argument,
3. the service quality/weak-user-as-victim counter argument, and
4. the human rights offence oriented counter argument.

We will start with presenting a short overview of the four generic counter arguments.

1. The unemployment of the staff counter argument

This argument seldom stands alone. Often it is combined with other arguments, for example, regarding negative effects on service quality and similar. Basically, the argument assumes that open libraries with prolonged opening hours will reduce the demand for professional and clerical staff, which eventually can lead to increased unemployment.

2. The economic/budget cut counter argument

In short, the budget cut argument states, that the main reason or the hidden purpose for implementing staff-less library practices, is to downgrade library provision to obtain budget cuts and to reduce library expenditures. The particularly mean, or smart, trick of the staff-less library model as a saving tool, is that it often, but quite unjustified according to the critiques, appears as if the open library model, by its widening of the number of opening hours, represents a user friendly service quality improvement.

3. The service quality/weak-user-as-victim counter argument

The third, generic open library counter argument, emphasizes that a number of the key quality parameters and values of a public library are bound to staff intensive activities such as communication, dialog, and interaction with the professional staff. Staff-less libraries, here, therefore, contribute to deprofessionalize library services because they do not allow fertile dialogs and fruitful interactions between professionals and users to take place during the staff-less opening hours. Especially, the weaker—in terms of social, economic, educational, and cultural resources—users, who really need guidance and advices from the professional staff are believed to suffer from the spread of staff-less opening hours.

4. The human rights offence oriented counter argument

The human rights argument claims that open libraries are contrary to human rights in some particular sense. A British critic, for example, claims that open libraries, by restricting access to people from 18 years old, are against the principle of equality by "denying young people the right to use the library independently" (Publiclibrarynews.com). The last word "independently" is important because young people and even children are usually also allowed to use the open libraries if they are accompanied by adults of, for example, plus 18 years.

Critical Voices

We will now take a closer look at some of the critical voices raised against the open library concept, illustrated by examples from different countries. First, however, we will consider an alternative collection of open library pros and cons compiled by the British library site, newpubliclibrariesnews.com.

To provide a fair presentation both the pros and cons, identified in the site, will be presented.

The pros of the open library, as seen by public library news, are the following:

Pros

- Avoid need for staffing, therefore, far cheaper per hour open, and so can extend opening hours far beyond what budgets would normally provide.
- Impersonal (some people prefer this).
- Hours can be set to suit the library, as can other restrictions. CCTV and remotely controlled equipment means security is provided.
- A good news story if (as is often the case) Open+ is twinned with cuts to libraries elsewhere.
- Regular users out of hours may police library themselves and assist those who need help.

http://www.publiclibrariesnews.com/about-public-libraries-news/unstaffed-librarie.

Upon closer inspection, however, some of the pros appear a bit strange. Why has, for example, the obvious pro argument that the prolonged opening hours enable users with long or odd working hours to use the library been omitted? And is it fair to suggest that people who appreciate open libraries are somewhat abnormal by adding in parenthesis "(some people prefer this)" (impersonal service). Indeed, there is no evidence supporting the point of view that people who appreciate the open library, typically, lack social competencies and take up hostile attitudes toward the staff. On the contrary, as will be demonstrated later, there seems to be no contradiction between being supporter of open libraries and—at the same time—being totally content and even excited about the services of the staff. Also the other so-called pros often seem to represent a more or less skewed picture as if the purpose has been to present the open library case as questionable as possible.

The so-called pros (mentioned earlier) are, however, also accompanied by a further quote from a report on the Peterborough experiences, which brings a bit more balance into the picture.

> *The council believes that the model that has been implemented offers people an enhanced service, without the council having to close any of the libraries. At a time when many other councils are having to close libraries, we are delighted that we have managed to develop a service that meets the needs of our residents whilst making the savings that we have to continue meeting the needs of the whole city.*
>
> *Peterborough Libraries (Vivacity).*

Furthermore, the references to literature on the site also contain sources with a more positive attitude toward the staff-less library concept. We will now take a look at the cons.

Cons

- Absence of personal touch, although remote conversations can be had.
- Equality issues: pilots show that far more men than women are willing to enter an apparently unsupervised library (ratio of 90% male to 10% female mentioned in one workshop attended by the author).
- Sense of "big brother is watching you" with CCTV.
- Question of access for those unaccompanied under 16 needs addressing in each authority. Barnet has decided 15 is the correct age.
- Cost: the system is cheaper than paid staffing but costs (especially set up) are still substantial. Cannot be used by those without library cards already.
- Not suitable for anyone with any special need. "Physical access will be difficult for anyone with such needs, and once in the library, without staff present, many disabled residents, perhaps with autism, or similar difficulties, will struggle to use the facilities provided..."
- Danger that regular library users during outside open hours will take ownership of the library, policing who is allowed in themselves. Would such users be checked by the authority to ensure they do not take advantage of other users during unstaffed times?
- "Thin end of the wedge": system is introduced to extend opening hours but, with cuts, may increasingly replace core hours, being used as an excuse or placebo for reducing opening hours.
- Dependent on IT. If there is a problem with the library servers or other computer-related issue, including servers or data, then the system does not work and the library has to close.

http://www.publiclibrariesnews.com/about-public-libraries-news/unstaffed-librarie.

Also, the cons require some comments.

The remarks about the lack of "personal touch" and "remote conversations" apparently refer to the presence of hot line telephones to a remote librarian. Experiences from the Nordic countries have shown that especially this help facility has only been used to a very limited extent.

Behind the second objection lies the fact that not only pilot studies but also data from operating open library systems have shown that open libraries in some cases, but not in all, seem to attract more men than women. The ratio is, however, seldom as extreme as the 90% men and 10% women mentioned. Typically, the Danish figures speak of about 60–70% men. The question is, furthermore, if this male dominance is caused by fear of entering an unsupervised library—as suggested—or by other reasons. Recent Danish data from Langeland showing quantitative male dominance also suggest that most of the visitors are elderly or old, which might weaken the fear hypothesis a bit. It is also questionable whether or not the occurrence of a relatively seldom male dominance in a public library is a weakness rather than the opposite. Among the intentions behind the open libraries were often an intention to attract new user groups. If it is true that the open library model attracts more men than women, the library might have gained new users and thereby fulfilled an important goal.

The big-brother-is-watching-you argument is also real. In Sweden, for example, the access for libraries to use video surveillance cameras is, therefore, for the same reasons, restricted. However, this open library counter argument is weakened by the fact that many people actually feel more safe when a site is monitored.

As to the counter argument about age restrictions it is added in an obtained e-mail response from the company Bibliotheca that age restrictions depend on local decisions. The restrictions are not at all required by the system. It is also mentioned implicitly than the user under the minimum age also can enter the library if they are accompanied by older people.

The point about costs also makes good sense—especially the one-off capital costs can be considerable. The same is true about the access problems for people with certain physical handicaps. Many libraries in Denmark and in other countries, however, which can be accessed in the staffed opening hours by users with wheelchairs, will probably also be accessible during the unstaffed hours.

The counter arguments also refer to as a danger that users take ownership and begin to police other users. Often ownership by users is considered to be a positive event but of course it is a problem when some users begin to keep order in the open library as if they were police. According to the experiences with open libraries in many countries such misuses, however, have not been reported often.

The last cons about the dependence on IT are highly relevant but not restricted to open library contexts. Also virtual library services and even staffed services are likely to be affected by IT problems or worse IT collapse.

After having looked upon this overview of open library pros and cons we will now focus on a number of critical positions in different countries.

National Variations—Ireland: Appealing to the Users

The most radical resistance is probably found in Ireland (www.impact.ie/impact-describes-sligo-staffless-library-plan-r). Here, the librarian's union, in 2014, has criticized plans to develop a staff-less library in Sligo county. The union describes staff-less library plans as a "retrograde step." Besides, the union has proclaimed that its members will not be assisting with the implementation of the planned staff-less library project. Furthermore, the staff has produced and distributed an information leaflet to the library users, to inform them about their concerns. The leaflet argues that the proposal to run staff-less opening hours will "erode the quality" of the library service by reducing the staffed opening hours. The distributed leaflet, furthermore, encourages library users to inform their local politicians, that a "library staffed by experienced and helpful staff cannot be replaced by an empty building with automated checkouts."

United Kingdom

Concerning the United Kingdom, we have already looked at the site, publiclibrarynews.com, which actually contains both positive and negative viewpoints. A closer examination of the character of the arguments presented, however, indicates that the position of the site is predominantly critical. The Peterborough case will be presented later (Hitchcock, 2016).

Denmark and Sweden

Typically, the resistance from unions and librarians is, however, much less offensive than seen in Ireland. The former Danish library union leader, Pernille Drost (2013) was certainly also skeptical toward the open libraries; but she was content to emphasize the importance of communication and guiding and to stress that libraries are more than books and shelves. Her members were neither called on to go on a strike nor to avoid assisting staff-less library implementation such as their Irish colleagues.

The Swede, Per Hjalmarson's, Critique

Another typical example of a combined budget cut and service quality critique is found in the critique of the Swedish librarian, Per Hjalmarson (2015b). He points out that the public sector share of the total economy has been reduced from 1990 to 2010. He also mentions, that about five to six branch libraries in Göteborg in Sweden have been closed within the last 25 years. He then concludes that the more-opened library model represents a new concept within a long tradition of downsizing library expenditures by reducing the staff costs.

Hjalmarson's key arguments are interesting considering the fact that open libraries often have been, at least in Denmark and in the United Kingdom, introduced to counteract branch library closures (see later the Aalborg case and the Peterborough case). Hjalmarson (2015a) is also shocked that members of the library staff have defended staff-less libraries, a behavior, which he sees as a "very dangerous trend." Finally, a social dimension is present in Hjalmarson's critique, assuming that especially the socially and economic weak users will suffer from the spread of the staff-less library concept. Besides, Hjalmarson makes the librarians' employment regards rather explicit: staff-less libraries are likely to increase unemployment among librarians. It appears that discussions about staff-less libraries are sometimes attempted to become a trade union issue. The Swedish discussion is interesting in itself. It is, however, also remarkable because it is often claimed by Swedish commentators that staff-less libraries in Sweden are much less associated with budget cuts than in Denmark (Engström, 2015).

An Effort to Reach a Balanced Point of View

An attempt to reach a balanced point of view as to staff-less libraries could be represented by an editorial of the former chair of the Danish Library Association, Vagn Ytte Larsen. He (Ytte Larsen, 2012) emphasizes that the underlying context and the basic conditions for implementation of the open library are significant. The optimal context, according to Ytte Larsen, occurs, when a 24/7 oriented solution is implemented together with increased staffed opening hours. An acceptable, but not optimal, situation is when a library, to avoid closing down one or more branch libraries, establishes an open library solution in the threatened branch(es). Not acceptable are solutions, where the staffed opening hours totally disappear.

Few Skeptical Library Leaders

The staff-less model has, in several countries, been recommended in national library development plans. In Denmark, for example, the latest national report, *Public libraries in the knowledge society* [*Folkebibliotekerne i vidensamfundet*] (2010, pp. 12–13), recommends that open libraries are used to revitalize smaller local library units and to increase library usage. Public library director, Jens Lauridsen, however, expresses a skeptical attitude toward the staff-less library concept. He points out that the open library reduces the library to become only a framework around a collection. In contrast, he describes an ideal dynamic and vibrant public library with a lot of activities such as reading aloud, learning and teaching, debates, etc.—all activities that require an interaction between the professional staff, external partners, and the user. Therefore, he clearly prefers extended staffed opening hours to a 24/7 framework with a high percentage of unstaffed opening hours. There are also examples of library leaders who have criticized concrete implementations without criticizing the open library concept as such (see the Ikast-Brande case (Chapter 4) and Mønsted, 2012).

Budget Cuts Have Occurred in Denmark but Less in Sweden

As to the financial oriented, budget cut argument, there is no doubt that many open library establishments, at least in Denmark and the United Kingdom, are related to contexts where library expenditures have been reduced. When I made the first open library mapping survey in this country in 2012 (Johannsen, 2012a), one specific question in the questionnaire, sent out to all Danish library directors, was, in fact, particularly devoted to identify and count the number and percentage of budget cut contexts. Here, about one-third of the respondents who then had established an open library solution answered that it had been implemented to obtain savings.

In Sweden, on the other hand, it was an important condition, not only for the staff but also for the users, that the more-opened library establishments were not linked to budget cuts, downsizing and firings of staff.

Library Users Typically Both Appreciate Their Staffs and Staff-Less Opening Hours

Indeed, it is a peculiar aspect of most user surveys concerning open libraries, that users, at the same time, emphasize that that they very much appreciate their librarians and that they nevertheless also express great satisfaction with the opportunities created by prolonged, unmanned opening hours. Indeed, user surveys can be very difficult to understand and interpret correctly. You can easily find user surveys that indicate that, for example, Danish library users today demand different services from their librarians than they did a few years ago and, in particular, before the Internet was introduced 20 years ago.

Changing Needs for Service

Broe Sørensen (2013), for example, points out, that information seeking assistance and help for technical problems have become less important services delivered by the staff. Nevertheless, it is also very difficult to spot new surveys documenting that public library users, in general, are dissatisfied with their staffs or that they think that they can completely do without the assistance of staffs. On the other hand, it is also very difficult to find surveys that document that no or only very few users do appreciate the new open, staff-less libraries.

Different Values

So it seems as if the qualities and benefits of staffed and prolonged unstaffed opening hours concerns different values. One important value associated with the open library is that you can use it when it fits you, whereas in the staffed opening hours you have access to personal assistance if you need it. But it is important to emphasize that the individual benefits gained—convenience and assistance—are quite different. Also from a broader societal point of view it is obvious that different values play a role, which the following case may illustrate.

Savings Versus Different Priorities

The relationship between budget cuts and the establishment or expansion of staff-less opening hours is often more complex than you think. There are, of course, economic aspects—both in terms of savings and costs—the latter associated with the establishment and operation of an open library. In the 1990s, loan and delivery automates transformed central clerical library processes into self-service functions. Also then economy and financial considerations played a role. Anyway, this transformation was considered to be legitimate by most users, although it was motivated by cost reduction purposes too. The essential point is, however, that cost reduction is not the only aspect associated with open libraries. At the individual level, we saw that values such as convenience and flexibility were involved and at a societal level the library's role in the local community is also influenced by prolonged opening hours. It is, for example, difficult to become an appreciated local meeting place with only 5–10 weekly opening hours, although they are all staffed.

The discussion whether it is preferable to save public money by closing down a number of branch libraries or by maintaining the existing branches as partially unstaffed units is not entirely an economic debate but as much a debate on different values and priorities. Three examples, a Danish public library case, Aalborg 2010 (Lerche, 2011), and a corresponding English case (Hitchcock, 2016), might illustrate my point. Finally, a third, an American case, Yacolt 2012 (Wallin, 2015), concerning a similar issue will be highlighted.

The Danish Aalborg 2010 Open Library Case

In Aalborg, a public library in a 200,000 inhabitants' municipality in Northern Jutland was directed to spare around six million Danish crowns (about £600,000) around 2010. The library leadership in Aalborg had decided upon a plan that would lead to closing down several branch libraries to maintain an appropriate quality standard in the remaining branches. However, the local politicians in Aalborg had different plans. To protect and stimulate a number of threatened or vulnerable local communities, they preferred a solution involving the preservation of a number of library units as part time open branch libraries.

Although the staffed opening hours of the chosen libraries were reduced from 40 to 23 h per week, the selected units and their associated cultural and local community values were believed to prosper from that solution. The alternative, to close down a local library, was considered by the local politicians to be more harmful to the preservation of the threatened local communities than the chosen staff-less library solution.

The Importance of Local Libraries in Depopulated Communities

Indeed, since the 1980s, public libraries in small rural communities have been an element in many planner's projects to save the local communities. It is, however, interesting to see how the open library concept, fundamentally, has changed the particular role of libraries in revitalizing local communities, threatened by depopulation. In the 1980s, cultural sociologists usually thought that library branches, located and integrated into local grocery stores as an additional function, were a solution to both depopulation and the widespread closures of local shops (Planstyrelsen, 198?). In some cases, it was thought that the shop staff should offer the library's services, whereas in other the job was to be done by librarians.

Libraries Can Play a Central Role

Here, the role of open libraries in saving threatened local communities today seems to be different. The staff-less library local community preservation strategy intends a much more central and independent role for the library. The staff-less library model allows the local library to continue as a place to stay and not only as a function to perform at a local shop. Impressions from a visit in 2011 can illustrate my point.

Vamdrup a Dark December Morning

When I visited Vamdrup, a 5000 inhabitants railway town in Jutland, I experienced a dark and misty December morning, where it was the local open library that represented the only, indoor, lighted, noncommercial place with public access (Johannsen, 2012a). Like a shining light in the darkness, the public library was here placed in a much more central position than as a subordinate, an additional function in a local super market or grocery shop, a destiny that was included in the visions of cultural sociologists of the 1980s. It should be mentioned here that it was not primarily the lack of imagination that prevented the 1980s sociologists to capture the new library role, but rather the stage of technological development that still had not experienced the opportunities of the future RFID and library automation technology to take place.

This Aalborg case, certainly, represented an agenda where budget reductions were in play; however, it seems to be much less certain that behind the initiatives of the local politicians were hidden an intention to downgrade the local library services, rather on the contrary. Reasonably, it was the priorities of local politicians and library professionals that differed. Indeed, the result in terms of spared money was the same, six million Danish crowns (about £700,000), in the initial plan of the librarians and in the resulting, open library solution.

Competing Cultural Policy Agendas

The important point to notice here is therefore that it can be somewhat misleading to use the budget cut critique because it is more often a question of different priorities than it is a pure budget cut issue. Indeed, the issue in Aalborg consisted of two different, competing political and cultural agendas. On the one hand, the professional library focusing on service quality and therefore preferring a solution with fewer, well-equipped branches to more, but mediocre local library units. On the other, the local politicians and—probably—also the local users, emphasizing the importance of protecting threatened communities throughout the preservation of local public institutions.

The English Peterborough 2016 Open Library Case

A similar story (Hitchcock, 2016) took place in Peterborough in the United Kingdom some years later in 2015–2016. Peterborough is a city of about 180,000 inhabitants (2011) in Cambridgeshire. In 2015, an amount of £25 million should be saved. The libraries should carry £400,000. A number of solutions were discussed by the city council, including a solution where all branches except the main library were closed down.

The Peterborough User Survey

Then a user survey was conducted with 5000 respondents among the library members. The results were striking. In addition, when asked "What would make you use your library more?" 75% of respondents stated that access to outside of normal opening hours was important. Here, open library technology helped to save council libraries. Bibliotheca's *open+* system was chosen and through the application of the system the required savings were met, all 10 branch libraries kept going and extended opening hours were implemented. Certainly, additional costs were added, both as one-off capital costs of £170,000 and as yearly costs per library of £1500. The required savings were obtained through saved salaries, which meant that the number of staffed opening hours was reduced.

Peterborough is one of the fastest growing cities in the United Kingdom. The City Council's library network, which is run by the independent trust, Vivacity, currently comprises 10 libraries and a mobile service across the city center and rural areas. This story tells about how Peterborough City Council overcame the challenge of budget cuts and avoided library closures with the implementation of open+.

The Library Was Highly Valued by Peterborough's Residents

Peterborough's residents truly value their local libraries: aside from borrowing items, they also use them as a community meeting place, an environment for learning and education, for computer and Internet access, and to participate in group activities such as Knit and Natter, Rhyme-time, and Storytime. With such a growth of usage, the Council felt it was important to evolve the library service for the future.

Following significant expenditure cuts, a review of the library service began in 2014, which led to a public consultation to understand how residents use, or would like to use, libraries. Lisa Roberts, Strategic Client Manager: Culture and Leisure, comments about the inevitable need for library transformation. She explains

> *The way our patrons use the library is changing: 90% of book loans are now completed through self-service kiosks, and staff now focus their time supporting people with more specific enquiries,*

The public consultation revealed that the most valued aspects of libraries are the ability to borrow books, access to information and location. The Council was passionate about finding a library model for the future whereby it would not have to compromise on service access and choice. Lisa recalls, "We spoke to Bibliotheca about open+ and thought 'This is something new and we'd like to explore it as different way forward'." After considering the more extreme alternatives, which all would have resulted in leaving local library users at a disadvantage, they chose open+. This option allowed all branches to remain open through a combination of staffed and extended self-service hours, including new Saturday opening hours, while still operating within budget constraints.

With the chosen open library solution, libraries can decide the days and hours they open for their communities to supplement access outside of the times when staff and volunteers are present. The system was an extension of Bibliotheca's current RFID system. In addition to managing opening hours and accessibility, open+ can be configured to control other elements of the library such as lights, public announcements, PCs, self-service kiosks, and a full CCTV surveillance system for a safe environment.

No Incidents in Peterborough

Since the initial implementation, more than 1000 members have opted in, with a peak usage of 250 open library users at the Central Library in 1 day. Lisa was pleased to report no major issues. We noticed this during the early stages of the pilot and used it as an opportunity to encourage more members to sign up.

Naturally, such an innovative way of managing libraries did not come without reservations. The main concern of the stakeholders' was the issue of safety and the "what if?" worst case scenarios that come to mind when contemplating public access in an unsupervised building. However, the reassurance of comprehensive risk assessments, consulting with the Safety Advisory Board, and the peace of mind that comes from the solution being a proven success in other libraries across the country and overseas helped to alleviate some of those concerns. The Council's Health and Safety Team approved open+ but a decision was made not to allow access to children under 16 unless accompanied by an adult during unstaffed hours.

To date, open+ is operational in four of the city's library sites, with the remaining installations due to be completed this summer. Lisa was thrilled to report positive feedback:

> *Customers have found it really easy! They are grateful that we have managed to keep our libraries open and for longer, and people are using them. The library is more alive now than it ever was before, with groups independently using the space. They are not restricted and they appreciate the choice that the new system offers.*

The library service now looks forward to widening community use to create multiuser sites. www.bibliotheca.com/open+.

Interestingly, Peterborough's Police Officers and Police Community Support Officers have all been issued with library cards. The move comes as community groups are encouraged to utilize libraries as an alternative workspace, which serve as a convenient base in between appointments. The invitation is extended to other organizations, and Lisa Roberts explained the added benefits that this brings: "the more people that use open+, the safer it feels with people popping in and out throughout the day. It's about establishing 'normal' again after this big transformation."

Another advantage is that staff members can be better utilized and distributed throughout the library network to undertake more customer-focused duties across fluctuated staffed opening hours.

A decision was also made to recruit four assistants to maintain a presence in four library sites where it was felt that extra safeguarding measures were needed during unstaffed hours. These assistants also carry out duties such as shelving and order processing so that librarians can focus on customer needs during staffed hours, adding more value to the library service. The open library solution has enabled Peterborough's libraries to overcome the challenge of transformation in times of austerity. Lisa Roberts advices other libraries that are considering implementing an open library concept:

> *I recommend that you try it out. It's not tangible until you see it in action. Talk it through with Bibliotheca; every library is different and it's not 'one size fits all', they will help you tailor the solution to your needs.*

As the Aalborg case, the Peterborough case deals with different priorities. Compared to the Aalborg case, the users play a more significant role in Peterborough through the conducted survey than in Aalborg. In both cases, however, the role of a nearby public library is important. We will now consider the third case, across the Atlantic, from the United States.

The American Yacolt 2012 Case

It is in Denmark and the United Kingdom that the budget cut argument against open libraries has been tried. Also, an example from the Washington state proves that community protection and open libraries are connected properties (Wallin, 2015). Yacolt is a town with about 4000 people living within a few miles. In 2012, it was decided to stop book mobile service in the county; a solution useful to the residents should therefore be found. Initially, it was planned to replace the book mobile service by a library in a city owned building, staffed 2 days per week for about 12 h. The problem with this solution was that the time when staff could most easily go there was also the time when many community members could not go there, because a lot of adult commuters left the town during the day to work elsewhere. Therefore, a staff-less library solution with an access control system at the front door, and four security cameras were chosen. The library, hereafter, was accessible to patrons 68 h per week, with hours available all 7 days. Staff members are in place at the library 5 days per week for about 3 h each day. The great difference to the initially planned time schedule was that only 2 days were "official" posted hours. The other days were more flexible and allowed staff to adapt to changing schedules. Special projects coordinator, Sam Wallin, concludes his story about Yacolt:

> *While the people of Yacolt haven't given up on having a larger, fully staffed library of their own, there is one thing they can get with an unstaffed library that would be very hard for the library district to deliver in Yacolt with a fully staffed library: 68 open hours per week over seven days.*
>
> *Wallin (2015, p. 8).*

Although the context of the American Yacolt case is somewhat different from, for example, the Danish Aalborg, and in particular the English Peterborough case, there are also conspicuous similarities: a phase dominated by traditional thinking (12 staffed hours per week) followed by a phase where local community interests and conditions are more included. Also Wallin's remarks that although some may view an unstaffed library as a way to cut costs, he believes it is

> *a viable model for purposeful expansion into low-service areas of a library district.*
>
> *Wallin (2015, p. 14).*

The staff-less solution makes it exactly possible to reduce the ongoing operating costs of opening a new facility.

Service Quality Counterargument

The second main type of critique points out that essential library service elements, typically provided by the staff, are missing in the staff-less model, in particular, the dialog between the user and the professional librarian, for example, about reference and information retrieval issues. Also, the more casual but fruitful user-librarian talks about the selection of fiction books and other library materials for work-related and leisure purposes could be mentioned. To many library users, opportunities to book talks with staff member without doubt are both appreciated and important and occurring in public libraries on a larger scale.

A radical and gritty version of the service quality oriented critique was expressed by an English user, Alan Wylie, February 22, 2016. He points out

> *I could put an aspirin in a tin outside my home and call it a hospital but it wouldn't be just like a library without a staff isn't a library, it's all spin and double speak (Publiclibrarynews.uk) (16 September, 2016).*

However, the emergence of the Internet and googling as a prevalent type of literature and information searching has certainly made many users feel—long before the staff-less library concept was "invented" in 2004—that the traditional information seeking assistances from a professional librarian had become more or less superfluous. Although it is true that everything is not on the net and available for googling, the Internet era has definitely not strengthened the prestige and authority of the professional librarian at least in business contexts, but probably also in public libraries (Johannsen, 2016). Nevertheless, his or her professional authority is certainly, not lost forever. Different efforts, for example, tailored courses in social media usage and the likes have apparently contributed to restore the librarians' professional prestige. Recent user surveys also point out that inspirational interaction and dialog with library staff members is still valued by the user (Broe Sørensen, 2013).

Human Rights Offence Counterargument

Some open library opponents in different countries, for example, the United Kingdom and Finland, maintain that open libraries represent a violation of some basic human rights. It means that not only the legitimacy of the open libraries is questioned but also their legality. It has, however, not yet been specified which particular national or international laws or conventions have been violated.

Summary

To summarize the pros and cons of open libraries is a complex issue for several reasons. First, principal participants and stakeholders have often quite different basic attitudes, values, and opinions: the users are mostly satisfied (we will return to the issue in more detail in Chapter 4), local authorities and government agencies are also excited, some library interest organizations are more lukewarm supporters and trade unions for librarians often directly hostile and dismissive. Among library directors in countries with longer traditions as to open libraries, Denmark for example, their attitudes are positive, in general, although stiff opponents can also be found here and there.

Second, another source of complexity appears when considering the arguments in favor of or against open libraries. Although the critical point of view, that behind all staff-less library projects lie hidden budget cut ambitions, is certainly false in many cases, the discussion whether restricted library resources should be used for either a little more staffed opening hours or for significantly more staff-less is fundamentally a question of priorities and values. The benefit of having access to the restful physical space on a late Saturday evening is also difficult to directly compare to the benefit of having a helpful librarian at hand during the conventional opening hours, especially, if the same user is occupied by shopping, gardening, jogging, and other important activities during the staffed opening hours.

Third, there are important societal issues involved, especially, concerning the maintenance and protected of threatened local communities. Sometimes such local community protecting concerns can be at odds with professional service ideals. Of course, user opinions and preferences are supposed to play a significant role a library planning. However, concerning open libraries, user surveys can be different to interpret because users at the same time apparently both appreciate the services of the staffs and prolonged, appropriate opening hours.

Although, values and political priorities play a significant role in the open library discussions, there are, nevertheless, a number of facts that should be respected by both the supporters and the opponents of the open libraries.

Perspective—Discourse Analyses as a Tool

Indeed, it can sometimes be rather complicated to compare and evaluate the validity of the different open library's pros and cons. Two Danes, a library researcher, Nanna Kann-Rasmussen and a library consultant, Elsebeth Tank (2016), have—inspired by an approach developed by the Australian political scientist, Carol Bacchi, called the WPR (What's the Problem Represented to be?), analyzed open library policies in Denmark.

In their analysis of the prevailing open library policy in Denmark, Kann-Rasmussen and Tank have added some further dimensions to Bacchi's WPR framework, for example, about who the addressed audiences are. They start with identifying the central sources and documents and the organization behind the policy they want to rely their examination on. Then they interpret how the problem is represented in central, government policy documents. Here, they find that availability constitutes a key element in the understanding of the problem by the central, library authorities in Denmark.

The question about how we can ensure that the availability of public libraries with few weekly opening hours in sparsely populated areas is secured has, apparently, become a key cultural policy issue during the so-called problem representation process. A particularly interesting aspect of this process is to identify which aspects that were not mentioned or made part of the discourse. Here, Kann-Rasmussen and Tank point out that the quality standards of communication and service belong to the not mentioned aspects of the open library discussions. Also, the value of the input provided by professional librarians was, apparently, also understated for the benefit of availability. Kann-Rasmussen and Tank conclude that "today, there is virtually no resistance against open libraries." Indeed, the power of the availability agenda has not been seriously challenged, disturbed, or substituted. This conclusion may conflict with the understanding presented earlier about widespread resistance among professionals toward the open library concept. However, both in Denmark and Sweden we can observe that resistance both among users and staff seems to disappear more or less when the open library projects are implemented that supports the theory of a dominating and unchallenged discourse.

Bacchi's WPR Method

From a methodological point of view, inspiration borrowed from the WPR approach can be useful when analyzing cultural policy issues such as open libraries. Indeed, based on an immediate consideration, the use of Bacchi's six WPR questions as a framework for policy analyses makes sense. The six WPR policy analysis questions are

1. What's the problem (e.g., of global warming, domestic violence, terrorism) represented to be in a specific policy?
2. What presuppositions and assumptions underlie this representation of the problem?
3. How has this representation of the problem come about?
4. What is left unproblematic in this problem representation? Where are the silences? Can the problem be thought about differently?
5. What effects are produced by this representation of the problem?
6. How/where has this representation of the problem been produced, disseminated, and defended? How could it be questioned, disrupted, and replaced? (Bacchi, 2009, p. xii)

Reservations Concerning Bacchi's Project

It is always useful and enlightening to consider the underlying background of public policies, including open libraries. However, the WPR approach becomes somewhat problematic when it aims to substitute other useful social science approaches. And it is exactly what Bacchi intends to do. In her 2009 book, *Analysing Policy: What's the Problem Represented To Be*, she emphasizes

> *The book has paid a good deal of attention to evidence-based policy approaches, due to their current near-hegemonic status in a wide range of policy areas, including health, criminal justice and education.*
>
> *Bacchi (2009, p. 252).*

Furthermore, she argues that

> *Social and political scientists bolster this image of "policy" as rational by offering "information" on which to base "knowledgeable governance ... In this understanding, policy analysts serve the role of "speaking truth to power" ... Evidence-based policy continues in this rationalist tradition which says that policy-makers make "good" decisions when they have "knowledge" ("evidence") "Knowledge in this view is decontextualized and sanitized "information"*

> *...A WPR approach, which puts in question the nature of "truth" and of social science "knowledge," is designed to return politics to policy analysis ... To this end I offer the approach as a replacement discourse for evidence-based policy, as developed below.*
>
> *Bacchi (2009, p. 251).*

The quote mentioned earlier, with its many inverted commas around words such as "information," "truth," and "knowledge," indicates a rather skeptical attitude toward traditional social science values, focusing on problem-solving, "what works?," and similar pragmatic problems. The idea—inspired by the French philosopher and historian, Michel Foucault (1926–84)—is instead to avoid to focus on the one correct response to an issue, but rather to examine how it is questioned, analyzed, classified, and regulated under specific, historical contexts. Bacchi explains

> *In order to perform this kind of assessment we need to direct attention to the effects that accompany specific problem representations ... this form of assessment does not refer to the standard policy approach to evaluation with focus on "outcomes." Indeed, ... a WPR approach puts in question the premise of evidence-based policy that lies behind such measurement approaches to evaluation. Effects as understood in a WPR approach, are much more subtle in their influence and rely for their understanding on a number of theoretical perspectives, including poststructuralist discourse psychology and feminist body theory*
>
> *Bacchi (2009, p. 15).*

Although it is difficult to say how "poststructuralist discourse psychology and feminist body theory" can contribute as regards to open libraries, Kann-Rasmussen and Tank conclude that availability constitutes a key element in the problem understanding of the central, library authorities in Denmark. Service quality, on the other hand, apparently plays only a minor role. What is interesting, however, is whether or not the application of "poststructuralist discourse psychology and feminist body theory" will change anything. The Danish and the Swedish library users, apparently, seem to be excited about prolonged, staff-less opening hours. But the Foucault-inspired thought, that many of the traditional elements in a public library—the open shelves, fiction literature, children's departments, reference departments, etc.—should not be seen as natural, taken-for-granted elements of a public library, although they are all historical and have—sometimes-been controversial features is still worth remembering. Indeed, we are often likely to forget that, and then it can be a good idea to be aware of the WBR-approach or similar methods.

We will now prepare a summary of the books eight basic research questions.

Research Questions

The main problems dealt with in the book are summarized in the following eight research questions. The questions will be, furthermore, explained and discussed in more details in the separate chapters as described later. Finally, the answers to the eight main questions are summed up in the Conclusions (Chapter 6). The chapter where the research question is primarily dealt with is indicated in the parentheses:

1. What characterizes the benefits and the challenges of the open public libraries? (Chapter 1)
2. Which particular aspects of the open libraries are controversial in professional discourses and what characterizes the pros and cons of the debate about open libraries? (Chapter 1)
3. How has the staff-less library concept developed and how is this development to be seen in a broader library history context? (Chapter 2)
4. How has the open library concept developed and spread in seven different countries in terms of primarily quantitative measures? (Chapter 2)
5. What characterizes the development of open library technological infrastructure? (Chapter 3)
6. What characterizes the themes and findings of reports and research about open libraries in seven different countries? (Chapter 3)
7. What characterizes the different reactions of the principal public library stakeholders (e.g., local politicians, users, the professional staff) upon the open library development? (Chapter 3)
8. Which role has local communities and the users played in the development of open libraries in different countries? (Chapter 4)
9. Which Critical Success Factors and associate strategies have until now been developed? (Chapter 5)
10. Which guidelines are useful when implementing an open library service? (Chapter 6)

The Basic Structure of the Book and Recommendations for Reading

Chapter 1 Presents an Overview and an Introduction

This book starts with an introduction to the topic covering both the present challenges and benefits and the professionally controversial issues associated with open libraries. The purpose is also give an overall introduction to the different aspects of open library model together with definitions and important concepts. If you only have time to read one chapter (about 40 pages), I would recommend chapter 1 both as an introduction to and an overview over the subject.

Chapters 2 and 3

The development of the concept within the public library sector, and compared to similar private and public sector examples, is dealt with in chapters 2 and 3. Chapter 2 describes other things and find an overview, including some key figures of the open library development in seven different countries, whereas chapter 3—also among other things—contains findings from reports with both quantitative and qualitative data.

Chapter 2

In general, the examples and points made will, primarily, concern open public libraries. When an example or pattern stems from academic library contexts, this connection will appear explicitly. Chapter 2 will also focus upon the ideas and purposes, which are found behind the self-service concepts developed within the private sector within the last 30–40 years; which similarities and differences there are compared to typical public library purposes and contexts? It seems, for example, as if digitalization and surveillance technology have played a much more significant role in public libraries, certainly, because self-service supermarkets most often are not staff-less because they both make use of checkout cashiers and other staff. Indeed, it seems as if the development of first the open shelves in libraries and later the open libraries has been quite revolutionary also compared to the private sector. However, it also seems as if the open library development also has been inspired by rather unexpected private businesses: tanning salons and banks. In the public sector, however, there are only very few examples comparable to the present day open libraries.

An important element of chapter 2 consists of an overview of staff-less libraries in seven different countries all over the world. The data for the overview are primarily collected directly by national library authorities and similar sources. The seven countries are listed under Chapter 3. As far as it has been possible to get the relevant information, the present number of staff-less libraries in the given country is stated, together with information about when and where in the country the first was opened. Because the official library statistic authorities in most cases have not yet begun to count the open libraries, some of the numbers are estimated and associated with uncertainty.

Chapter 3

Chapter 3 then deals with the history and evolution of open library solutions in primarily the public libraries. Especially, the chapter will focus on national and geographic differences, in particular, concerning a selected number of seven countries. Compared to the preceding chapter the evidence of chapter 3 stems primarily from reports and research articles in peer reviewed articles complemented with data from interviews, internet sources, and the like. The seven countries (in alphabetic order) selected are:

1. Denmark
2. Finland
3. Germany
4. Norway
5. Sweden
6. United Kingdom
7. United States

The selection made is primarily determined by languages intelligible (English, German, and Scandinavian languages) to the author. Where staff-less library experiences from other countries, for example, in Far East Asia, are available in sources in English or German such evidence is, of course, also referred to. Indeed, some very important developments have taken place in countries such as China, Japan, Singapore, and Taiwan. As the history of staff-less libraries is relatively short, the earliest examples stem from about 2004; the main focus of the chapter, consequently, will be the national differences and patterns.

The experiences from the seven chosen countries will be presented in a systematic manner. Chapter 2 will contain the basic data on, for example, the first open library in the country, the present number of open libraries, statistics if available, and similar information. Chapter 3 contains a review of existing reports in English, German, or the Nordic languages on open libraries. Information suitable for judging the coverage, reliability, and validity of the reports will be given here as far as they are available together with a summary of their main results. However, some of the figures and observations are also likely to be presented and dealt with in chapter 4 on local communities and on users and usage, and also in other chapters in relevant contexts.

Chapter 4

Chapter 4 deals with both the local communities and the users of the open libraries. Certainly, there are some typical first movers when discussing local community characteristics. However, does the apparent rural and suburban character of the first generation of staff-less libraries also implicate that this particular library type will not also fit into an urban or metropolitan context? Although we for natural reasons do not have particularly detailed pictures of the typical staff-less library users, it seems interesting to find out what distinguish staff-less library users compared to users of the staffed libraries or to what extent there, actually, are sociodemographic, life style oriented, and other differences between the two user categories. As to the eight research questions dealt with, focus is obviously here on question number eight. The chapter also includes a section on technology, prepared with valuable input from the two software companies, Bibliotheca and Cordura.

Chapter 5

Chapter 5, presents a number of open library strategies based upon four critical success factors. The strategies are (1) strategies to prevent vandalism, (2) strategies to make the library an inviting and attractive place to stay, (3) strategies to promote the library as a third place, and (4) strategies to improve communication and user-friendly-ness.

Chapter 6

The chapter contains recommendations and advices for the implementation of open library solution. Like in chapter 4, the two software companies, Bibliotheca and Cordura, have assisted with both valuable inputs and practical library cases from different countries.

Chapter 7

Finally, chapter 7 summarizes the main conclusions of the whole book.

Tables, Figures, and Boxes

In most of the chapters, tables, figures, and boxes will appear with relevant content from libraries in different contexts and surroundings. In many of the boxes librarians from Denmark, Norway, and Finland have been asked to contribute with open library case stories. In other boxes, case stories provided by Bibliotheca and Cordura will appear. In one box, an ongoing Ph.D.-project on open libraries is reported. I have in all the cases emphasized that the libraries involved could be identified through places and through names of employees, directors, and other sources. Lists of tables, figures, and boxes appear in the list of contents.

Open Library Service in a Broader Perspective

2

Library History Perspective—The Gradual Opening of the Library

A modern open or staff-less library is characterized by a number of features; the most important ones are:

- free access to the physical library up to 24/7 h a week,
- access to the book shelves,
- free access to the basic library services and transactions,
- free access to online catalogs,
- free access to sources on the net,
- free access to the library's computers.

In a library history perspective, however, the present openness is a result of a long historical development with a number of evolutionary steps and with significant national and historical variations. We will take a closer look at those steps to consider characteristic similarities and differences. Having the discussions in the previous chapter about pros and cons in mind, we will also consider whether the development toward the opening of the library should be considered to be a progress or the opposite.

Here, it is, to start with, important to repeat the distinction between two related but, nevertheless, somewhat different concepts:

1. Self-services
2. Staff-less or open library services

A self-service is primarily characterized by a practice, where the customer has direct access to the goods. He or she then selects what he wants and brings the chosen objects to a checkout desk or counter where the goods are either paid for (e.g., in supermarkets) or registered as on loan (in libraries). The difference, compared to staff-less services, is that staff members are usually found somewhere, either at the checkout desk or usually also available at other places in the self-service shops.

A modern public library is, also in its staffed opening hours, very much based on a self-service model. Indeed, you can use the library within the staffed hours, practically, without any contact at all with the professional librarians or with the clerical staff as well. This has been the situation in many public libraries since the 1990s, when the automation of the checkout and delivery functions was implemented.

The staff-less model is characterized—as the term indicates—by a different situation, where there is no staff available for assisting the user. Besides, the user is,

Staff-Less Libraries. http://dx.doi.org/10.1016/B978-0-08-101923-8.00002-2

furthermore, supposed to lock herself or himself into the library. In the public and especially in the private sector we also find a lot of self-service but very few staff-less solutions like the open library model. You may object that many public and private online services where you, for example, can buy train tickets or pay taxes are staff-less. It is correct. However, there are still very few public and private services outside transportation where you get unstaffed access to spaces and physical facilities. Tanning salons are among the few stationary private sector examples. In banks, the cash dispenser outside the bank is also staff-less. However, access to the physical spaces of the bank is not included.

Libraries have been through a number of development steps to achieve the present open library stage. I have divided them into three primary phases.

Open Access

About a little more than a 100 years ago, open access to the library's book shelves was quite unusual in public libraries. The "open access revolution", a term coined by library historian, Alistair Black, defines open access as the "freedom to browse and self-select in public library lending departments" (Black, 1994, p. 3). The revolution started in the United Kingdom in 1894 and some years earlier in the United States. In the United Kingdom, the Clerkenwell Public Library in North London was the first public library offering Open access to lending services. Many other public libraries followed over the next two decades.

The open access revolution forms the headline for a major step in the long history of the opening of the public libraries. There are both interesting similarities and differences to the later open libraries, which started about 110 years after the open access revolution in the United Kingdom—this time in a remote village in Denmark.

Democracy and Professional Prestige

We will now focus a little more on the open access revolution to compare it to the related, but still different, "open library revolution", which started a little more than a century afterward. Black (1994) has examined the open access revolution, which he sees as a turning point after the passive 1850–1894 era, where the position of the public libraries was more defensive and closed. According to Black, the revolution was characterized by an influential democratic impulse and imbued with "visionary, egalitarian idealism" aiming at a "consumer democracy" (Black, 1994, p. 3). However, Black also tries to consider the open access revolution as not merely a democratic initiative, but also as a power-driven "librarianship-derived impulse", aimed at raising the public library and the public librarians in the estimation of the masses. Open access also helped the library profession to develop a special prestigious scientific subdiscipline and a niche body of knowledge called "library science" (Black, 1994, p. 5).

This interpretation of what lies behind the open access revolution is inspired by the theories of the French philosopher and historian, Michel Foucault (1926–1984).

According to Foucault professional knowledge can never be based on notions like the neutrality of science; but on the contrary on "interested concerns like status and ideology" (Black, 1994, p. 7). This argument can also be applied to the open access revolution. However, I do not think that it is particularly surprising or revealing or in any way suspicious, that professionals try to increase and develop their own professional knowledge. Furthermore, it should be added that Foucault himself, never did mention libraries in his various deliberations on the relationship between power and knowledge (Black, 1994, p. 7).

Pros and Cons of Open Shelves

As mentioned, it is also interesting to compare the pros and cons of open access with the similar positions regarding the later open library revolution. Opponents of open access, in general, revolved around accusations of librarian self-interest. It was alleged that open access primarily would be adopted by librarians "ignorant of their calling or lazy, or both" (Black, 1994, p. 4). Open access was called a "help-yourself-and-do-not-bother-me" system. It was also reckoned that open access required twice as much space as closed access. One reason was the restriction in the height of shelving to ensure that users did not face the impractical task of using a ladder.

Among the followers of open access, the democratic aspects were often referred to. Also the improvements in signage and transparent classification—either after subject or alphabetic order—made open access a user friendly feature in public libraries. But also the more indirect effects of open access on creating a dialogue friendly atmosphere in the library were observed by simultaneous observers such as the Danish author, Harald Herdal (1900–1979) in his memories:

> *However, when you had become more confidential of the library premises during the evenings, and I had no other times to visit, a certain nice mood arose when the loaners went around the shelves searching, chatting with each other, seeking advice, recommending or deprecating a book they had just read ... here was easy access to the books according to the principle of "open shelves." You could pick and choose, find the books you liked (if they were not already lent) and if you knew what you wanted to read. That could be difficult ... There was also something called "closed shelves." To borrow books placed there, required that you must ask a librarian for permission beforehand.*
>
> *Translated by CGJ, Dahlkild (2011, p. 115).*

As to the second consequence, the need for more space, quite similar problems have emerged related to the establishment of open libraries, where a lot of seldom used library materials in Danish libraries have been discarded to create a user friendly and foreseeable library interior.

Self-service in the 1920s, meant that the user was allowed self to "graze" on the shelves and to select the titles he or she wanted, as the quotations above from Harald Herdal's memories illustrate. The online revolution of the 1990s and onward meant a new opening, because it allowed the user the freedom to access the catalog from home and to find out whether certain desired titles were out or available. The online catalog, furthermore, democratized the public libraries by opening up in many libraries access to information on the supply and demand of individual titles. Such knowledge about supply and demand of library material had so far been a professional privilege.

Early Open Shelves Libraries in the United States

We have until now only looked at open access in public libraries in the United Kingdom. However, the same development of open access also took place in approximately the same period—the decades around and after 1900—in North America and Scandinavia (Dahlkild, 2011, p. 36). However, it seems that library pioneers in England and the Scandinavian countries were inspired by very early open access public libraries in the United States (Black, 1994, p. 4). One of the first was opened in Rhode Island around 1890 (Dahlkild, 2011, pp. 36–37).

Open Shelves and Serendipity

In a historical perspective open shelves were not something, which could be taken for granted in public libraries. In Danish public libraries, the open shelves idea was introduced in earnest around 1909 (Dahlkild, 2015, p. 81). The new openness was legitimated through several reasons. First, it was considered important by public library authorities to allow and even stimulate communication and sharing of reading experiences between the users. Second, the open access to the shelves could facilitate the user's chance of finding something unexpected and useful at the same time. This specific form of benefit is now often named "serendipity". The library architectural historian, Nan Dahlkild, has reflected over the apparent paradox that although the interior open shelves, on the one hand, communicated democratic values and equality, the exterior of many public libraries between WW1 and WW2, on the other, appeared as knowledge temples with imposing, absolutist portals, and monumental buildings (Dahlkild, 2015, pp. 81–84).

Open Shelves Did Not Penetrate Public Libraries Everywhere

On the other hand, in countries, such as Germany and other Central European states, closed shelves remained commonplace till after the Second World War. A German book about public library architecture from 1930 reasoned that the English open access libraries "missed the opportunity to individual and qualitative immersion", which—in contrast—was supposed to characterize the German public libraries (Volksbuchereien) (Dahlkild, 2011, pp. 37–38).

Library Supervision

One interesting common issue of open access and open libraries concerns the concept "library supervision" and its application to surveillance architecture in open access libraries. Black remarks that "radial designs" of lending libraries became very common. Here, bookshelves positioned so that staff could see clearly down any aisle at any given moment. This architecture is said to be inspired by Jeremy Bentham's (1748–1832) prison design, where the supervisor could view inmates, but not vice versa (Black, 1994, p. 7).

Panoptic Design in Open Libraries

The issue is interesting because we find the same type of considerations related to open library architecture and interior design. The importance of an overview with no hidden or dark corners in the library has been emphasized by library planners, when transforming a staffed into an open library. However, whereas Bentham's prison design was not invented to the benefit of the prisoners, the panoptic interior design and the use of video surveillance of the open libraries were certainly features meant to increase the security of the user during the staff-less hours. Of course, in cases of vandalism, data from the video surveillance could be useful too; but the purposes of panoptic design and video surveillance were more than a mere Foucauldian control and power exercise.

Another aspect of the similarities and differences between open access and open libraries concerns who promoted the ideas. We find that librarians—at least in the United States, the United Kingdom, and Scandinavia—were among the supporters of open access. As to the open libraries more than a century later we find the same pattern. Both in Denmark and in Sweden were the first open libraries initiated and supported by professional librarians. Later, however, also local politicians began to support the open libraries and, sometimes, the professional librarians had different priorities as, for example, seen in the Danish Aalborg open library case.

Automated Self-delivery and Self-checkout of Library Materials

The next step occurred when the clerical functions associated with delivery and loan were automated from the 1990s and onwards. In 1991, the company 3M in collaboration with the Campbellton Library of the University of Western Sydney first tested an automated checkout system (Cookman & Burgess, 1996). In the 1990s, self-checkout service became widely used in libraries (Tseng & Kuo, 2009, p. 412). Such changes were not exclusive library developments but part of a broader trend of self-help in modern society (Tseng & Kuo, 2009, p. 413).

Open Libraries (The Open Library Revolution)

The world's first, fully developed open public library started in 2004 in a Danish village, Gjern, 16 km outside the provincial town, Silkeborg (Larsen, 2007). At about the same time, a similar experiment took place at a site very far from the sparsely populated, agricultural surroundings of Gjern. It was at the Taipei public library in Taiwan in 2006 (Tseng & Kuo, 2009). Even earlier a Do-It-Yourself library was opened in December 1, 2002—in Singapore. Why the later Gjern library, nevertheless, is considered to be the world's first open library will be explained below.

Far Eastern Experiences

Despite the considerable distance between the two public libraries in Gjern and in Taipei, a number of similarities were conspicuous. Both were based on radio frequency identification (RFID) technology combined with an automated checkout system. Tseng has characterized RFID as "one of the 10 most important inventions of the 21st century" (Tseng & Kuo, 2009, p. 411). Besides, RFID tags include an efficient antitheft system, which makes beeping sounds whenever materials, which have not been checked out correctly, leave the library. There were also differences. Although Silkeborg chose the word "self-service" for its service model, Taipei used the word "intelligent library". Also in two other Far Eastern countries, China and Singapore, experiments with intelligent staff-less libraries took place (Lek Choh, 2003; Tseng & Kuo, 2009; Wieldraaijer, 2005).

Early Prototypes From the 1990s Missed an Important Detail

Already, in the 1990s a prototype staff-less library was tested at subway stations in Shanghai and Tokyo. About 1996 Tokyo had set up 14 subway libraries, offering reading and borrowing services 24h a day. Compared to the decennium to follow, the Tokyo model, however, missed an essential feature. Users could, namely, borrow books without having to go through an automated checkout procedure. As most passengers failed to live up to the high moral requirements that such a system demanded, the book collections were quickly diminished. At one library, 200 volumes were borrowed on an average day, but only 10 were returned. In Shanghai, after 1 month, the library discovered that it had lost more than 10,000 items of its 40,000 books and magazines (Tseng & Kuo, 2009, pp. 414–415).

Despite or rather because of these disappointing results a new type with effective antitheft procedures, the "Open Book Ximen Intelligent Library" was established in the underground passageway of the Ximen Station in Taipei. One aspect of the antitheft procedures applied might, however, appear somewhat problematic. Tseng and Kuo (2009, p. 415) explain that an alarm goes off and the "gate remains closed", should a user try to leave the library with a non-checked out book. Although this procedure might be reasonable toward the sinner, it seems difficult to justify in proportion to the respectful users that are locked in as well.

The World's First Open Library: Gjern or Singapore?

Also, at another Far Eastern destination, in Singapore, open library experiments took place. Some authors like Johansson, Lindberg, and Eckerdal (2015) even claim with reference to Lek Choh (2003) that the Singapore library was the world's first open library. The Singapore "Totally DIY (Do-It-Yourself) library" was opened in December 1, 2002. It was 2 years before the Gjern branch library. According to Lek Choh (2003), it was run without any staff onsite. Technically, the Singapore library—like the Gjern—was based on RFID technology and tags. Their purposes were also similar. To make library visits "hassle-free, enjoyable and enriching" sounded the aims of the Singapore prototype

(Lek Choh, 2003, p. 2). A "cybrarian" kiosk through which the users of the totally DIY library could contact a librarian by telephone and co-browsing facilities were also elements present in both the Danish and the Singapore staff-less library. As to the 24/7 goal, there was probably a small difference between Gjern and Singapore. Extended opening hours were, apparently, not the most important purpose in Singapore.

Anyway, the Singapore library planners were also ready "to extend opening hours where there is a need" (Lek Choh, 2003, p. 5). Indeed, the physical library facilities and the library as a place to stay and to meet others were, maybe, less pronounced goals in the Singapore project. To bring library services to more people in a more accessible manner (Lek Choh, 2003, p. 5) was the overall purpose in Singapore. However, apart from the already mentioned differences and variations the two public library projects appeared to be related both in terms of purpose and applied technology. One important difference is, however, that the Singapore DIY library was located within a larger shopping mall, whereas access to the Gjern branch required a separate lock-in procedure to unlock the doors. One can, of course, discuss whether this particular lock-in procedure is that essential. In countries like the Nordic with libraries in small independent buildings such procedures are without doubt necessary, although in a densely populated City State with large shopping malls physical lock-in procedures are more dispensable. Then the conclusion is that the world's first open library with physical lock-in procedures was established in Gjern in Denmark in 2004.

Nordic and Far Eastern Open Library Experiences: Other Differences

What is also interesting, when comparing the first generation of staff-less libraries in Denmark and the Far East, are the quite different preferences as to location. In Denmark, the first and many of the following staff-less libraries were placed in thinly populated rural environments, whereas the Far Eastern library planners chose rather different sites for their intelligent libraries: busy, metropolitan neighborhoods. Later this situation has changed, at least in Denmark, where staff-less libraries are now found in urban environments, towns, and cities too.

The next sections will deal with the staff-less library development in seven selected countries. In the following Chapter 3, two sections will deal with the research-oriented literature (reports, peer-reviewed journal articles, and the like) about the development in the same seven countries).

Open Libraries in Seven Countries—Overview

The purpose of this section is to present an overview of the development of the open library model in seven different countries. The primary criteria for selecting a country for the list has, of course, been the occurrence of the open library type in the country. An additional criterion has been language. Certainly, cases of open libraries from Far

East countries (China, Japan, Singapore, and Taiwan) have been published in international journals and the relevant articles have also been referred to in this book but language problems have made it difficult to find out about more specific details. This reservation is also valid toward Netherlands where open library projects also have taken place. There are also countries, where both the staff-less concept has been tested or where interest in the concept has been showed and where the author can understand the language but which despite all that was not included in the list below—mostly because relevant cases have not been reported in the literature. Countries like, for example, Belgium, Canada, and Switzerland belong to that category.

Concerning the evolution of open libraries around the world, the figures below illustrate that the libraries are primarily found in the four Nordic countries and in the United Kingdom. As already mentioned, a number of very early staff-less libraries are also found in Far East Asia. In Table 2.1 below, an overview of the situation in seven chosen countries is presented.

Sources

It is rather complicated to find out how many staff-less libraries that are running today in different countries. The primary sources are:

- Central, national library agencies, statistical services, and library associations (Denmark, Norway, the Netherlands)
- Software suppliers (such as Axiell, Bibliotheca, Cordura, etc.) can provide data from different countries too
- Professional journals, online sites, and the like
- E-mail interviews, interviews, questionnaire surveys, and the like

One particular problem is that because the library type is quite new, it is often not yet contained in the categories of the official library statistic. Sometimes, like in Sweden, for example, the national library statistic authorities count the number of more-opened weekly opening hours but not the number of library units offering the more-opened services.

Table 2.1 Open Libraries in Seven Countries

Country (Year of First in Parenthesis)	
Denmark (2004)	297
Finland (2009/2012)	(90)
Germany (2014)	1
Norway (2013)	78
Sweden (2009)	(50)
United Kingdom (2013)	35
United States (2016)	?

Notes: A parenthesis around the number means that the number is estimated. In the later sections about the individual countries, the sources of information are specified.

We will now look at the available data from the nine countries. In each country the particular source or sources are mentioned. The countries are described in alphabetic order.

Denmark

The data are, primarily, based on my own research, for example, Johannsen (2012a), from online available authoritative sources and from software companies like Bibliotheca and Cordura.

Basic Country Data

Denmark has 5.7 million inhabitants (2016), an area of 42,924 km^2 and 98 municipalities. The country had (in 2014) 97 main libraries and 354 branch libraries (http://slks.dk).

First Open Library

The first open library in Denmark was established in 2004 in the small village Gjern in Jutland. It is a somewhat controversial issue whether the Gjern library or a library in Singapore (Lek Choh, 2003) represents the world's first open library. Johansson et al. (2015) claim with reference to Lek Choh (2003) that the Singapore library was the first. Mogens Larsen, one of the founding fathers of the Gjern library, however, has argued why his library—and not the one in Singapore—should be considered the world first open library (Box 2.1).

Statistics

According to the latest official statistics from the Danish Agency for Palaces and Culture, which now manages the public libraries in Denmark, the number of open libraries grew to 297 in 2016 (Slots-og kulturstyrelsen, 2016; www.slks.dk). The explosive growth of staff-less libraries in Denmark from 2009 to 2016 is illustrated by Fig. 2.1.

Only Five Totally Staff-Less Libraries in Denmark

Among the present open library units, only five are, according to a 2016 overview by the Danish Agency for libraries and media, totally staff-less, driven by volunteers.

Staffed and Unstaffed Opening Hours

In 2011, Johannsen (2012a, 2012b) calculated the distribution of staffed and unstaffed opening hours in the then 76 Danish public libraries, which in the summer 2011 offered open library access. The figures are presented in Table 2.2 below:

Table 2.2 represents data from a country where the open library idea, at least until now, has had its greatest impact. It should, however, be remembered that the Danish figures cannot, immediately, be compared to, for example, the Swedish, because the

Box 2.1 Gjern 2004—The World's First Open Library—Case

Mogens Larsen played a central role in the establishment—he tells:

"It started in 2004 when Silkeborg library was asked to take over the operation of the libraries in Gjern, while its leader then had got a new job and left.

Focus-group interviews with users in the municipality revealed a common wish to introduce prolonged opening hours and extended availability in the municipality's five libraries. I was then asked, using project funds from the National Library Authority, to deal with the wishes that the focus-group interviews had revealed. Here, availability was most important and inspired by banks, which both allowed access to the buildings and also offered an option where you could consult an advisor. At the same time, a cash point or an automated teller machine (ATM) outside the bank was a necessary service that complemented the service model wanted by the customers.

At that time, the libraries were World Champions in virtual services through prompt expedition of bookings through the inter-urban loan system. The loan system worked around the clock while the local library where the user picked up the ordered materials were only opened few hours a week. Self-service was therefore necessary to create a better service.

The users were also ready for self-service. However, through the focus-group interviews they emphasized that self-service should only be a supplement to the presence of the staffed and that signs should tell when the staff was present. User-friendliness was emphasized too—the technique should not be more difficult 'than the bottle vending machine in a super market'.

Self-service in libraries were then not at all common. Internationally, experiences from Singapore where self-service was implemented inside the library were available. But not concerning *access* to the library which was a key element of the Gjern project.

To me an adequate service concept consists of a mix including: common staffed opening hours, book-a-librarian services and modern information technology based self-service. I'm convinced that if we had implemented the concept earlier several branch libraries which had been closed down maybe were saved."

By Mogens Larsen, library developer.

Swedish figures also include numbers from libraries, which are staffed all the time. The Danish figures, on the contrary, only concern the 76 library units out of more than 300, that, in 2011, had one or more units with weekly staff-less opening hours. In 2011, more than half of Denmark's municipal library systems did not offer staff-less library services.

Vandalism

Vandalism and unrest are a rare but not totally unknown phenomenon in the many Danish open libraries. Both in the provincial town, Aalborg, and in the capital,

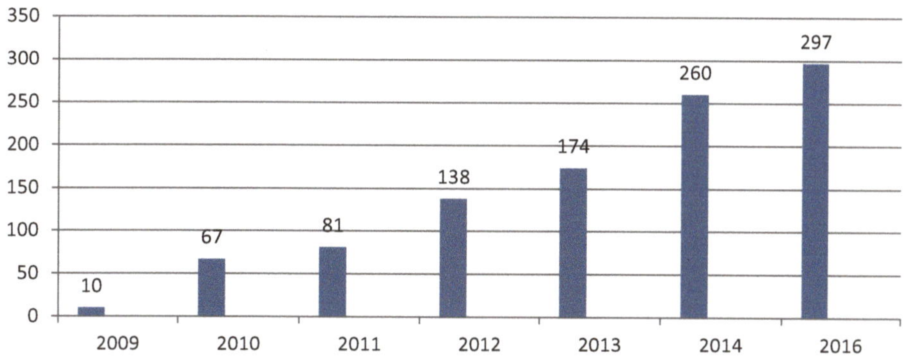

Figure 2.1 Open public libraries—Denmark 2009–2016.
Jørgensen, J. (2013). Selvbetjente biblioteker boomer [Self-service libraries are growing fast]. *Momentum*. http://www.kl.dk/Momentum/momentum2013-8-2-idl29736; Slots-og Kulturstyrelsen-Oversigt over åbne biblioteker pr. (September 1, 2016). www.slks.dk.

Table 2.2 Staffed and Unstaffed Opening Hours per Week in 76 Open Public Libraries in 2011 in Denmark—Unit: Hours per Week

	Hours per Week	Average Hours per Week	Percentage of Hours per Week (%)
Total staffed opening hours	1515	20	22
Total unstaffed opening hours	5281	69	78
Total staffed and unstaffed opening hours	6796	88	100

Johannsen, C.G. (2012a). *Åbne biblioteker—et kortlægningsprojekt [Open libraries—a mapping project]*. Copenhagen: Det Informationsvidenskabelige Akademi, Table 1.

Copenhagen, cases of vandalism and unrest have occurred in some but, by far, all the urban branch libraries of the two towns. In Chapter 5, a description of what can be done to prevent the problems will be presented. Not only in urban districts but also in rural, for example, Præstø, vandalism, and unrest have taken place.

Resistance

Resistance against opened libraries in Denmark is found, mostly, within the librarians' union where a number of slightly critical articles have been published (e.g., Lerche, 2010, 2012a, 2012b; Mønsted, 2012). Very few library leaders (Lauridsen, 2011) have expressed critical opinions. The Danish Library Association has welcomed the open libraries except in cases, where the staffed opening hours have totally disappeared in connection with the introduction of the open library model (Ytte Larsen, 2012).

Best Practice Libraries

In this book, Danish best practices examples are represented through accounts, in separate boxes, on the libraries in Bornholm, Copenhagen, Sønderborg, and Aalborg.

Trends

It is commonly assumed in the library sector, both by supporters and critiques, that the ongoing trend in favor of open libraries will continue.

We'll take a closer look on the available research reports on the Danish open libraries in the following Chapter 3.

Finland

The main source is here, responses received July 20, 2016, from senior officer (public libraries), AVI, Kristiina Kontiainen (KK), who very carefully has filled out my questionnaire concerning staff-less libraries in Finland, and complemented it with many highly relevant and valuable links and references. Here, only a short review of the history and the current numbers of staff-less libraries in Finland. A more detailed presentation, based on published reports of the Finnish staff-less libraries will be found in Chapter 3.

Basic Country Data

Finland had 5.4 million inhabitants, 301 municipalities, and 738 public library units (288 main and 450 branch libraries) in 2015. Finland is a country of 338,424 km^2.

First Open Library

The first staff-less library in Finland opened in 2009 in Ylitornio (Övertorneå in Swedish) library when it opened its newspaper room as a self-service department. The room was opened daily from 7 a.m. to 10 p.m. The age limit was 16 years. In Hämeenlinna branch library in Jukola the self-service hours began in 2012 in addition to the normal opening hours. If one choses to define just a part of a library as a self-service library, then the first staff-less library opened in 2009, but if one applies a more stringent criteria, the first entirely staff-less library in Finland opened in 2012.

Only One Totally Staff-Less Library

In Finland there is only one public library, which is open totally without staff. It is a little village library in Mynämäki municipality: Mietoisten itsepalvelukirjasto, self-service library in the village, Mietoinen, which was renewed and opened as totally self-service library in 2014.

All the other more-opened (Swedish: mer öppen) libraries in Finland are only partly staff-less. More and more libraries are open both with staff and without staff. Twenty percent of the municipalities have self-service hours in their libraries. During the last 10 years there have been lots of mergers of municipalities (kommunsammanslagningar). To avoid a total abolition of branch libraries (former main libraries), some libraries have found out ways to intensify their function by, for example, having staff-less opening hours. Based on the information that 20% of the municipalities had staff-less hours, we can calculate that there are at least 60 more-opened public libraries with open library branches in Finland. If we, furthermore, assume that 20% of the branch libraries are open libraries, we can estimate the total number of open branch libraries to be about 90 in Finland.

Statistics

From 2014 to 2015 the total number of staff-less opening hours in Finnish public libraries increased from 57,691 to 124,608, which corresponds to a growth of 116%. Compared to the total number of opening hours, which was 1,422,896 in 2014 and 1,385,784 in 2015, the proportion of unstaffed opening hours grew from 4% in 2014 to 9% in 2015.

For example, in the Southern Finland (incl. Helsinki area and five counties) the statistics 2015 show that 7.4% of all opening hours were self-service. In comparison with the year 2014, the increase of the self-service was 130%. Opening hours with staff decreased 4%. Without self-service hours, the total number of opening hours would have decreased.

As to the negative or positive effects of staff-less libraries on library usage in terms of visits and loans, KK, remarks that it is difficult to say how much the self-service has affected visits and loans. In general, both loans and visits have decreased in Finnish libraries. However, the number of visitors to library events has increased. Overall, she does not think that self-service has affected library statistics very much. If it shows up that customers spend more time in the library, because the self-service hours are more convenient than the "traditional" opening hours, it is more a question of quality than quantity.

Vandalism

Concerning cases of vandalism, KK confirms that also in Finland there have been less problems than was expected. She adds that the premises are usually under camera surveillance.

Resistance

Asked how the library profession and the users like the new staff-less concept, KK emphasizes that those who are in charge of financing clearly see the advantages of the effective use of the premises and collections; they see self-service as an extension of good service. Some professionals, however, are worried about the quality of service. Some fear that the self-service is merely a means of saving money by reducing the qualified staff. Because some libraries have age limits during the self-service hours (15–18 years), this particular aspect has caused a discussion about accessibility. KK supposes that the users do not want to have totally staff-less libraries. Customer service plus extended opening hours with self-service seem, however, to be a workable combination.

Best Practice Libraries

Concerning any Finnish examples of "best practices" or "lessons learned" concerning self-service libraries, KK refers to the Vantaa City Library (see the box below).

Trends

Concerning the long-term trends, KK assumes that the number of (at least partial) staff-less libraries will increase for several reason:

- the municipalities want to function more effectively,
- the state has supported libraries by financing open library projects, if the aim was to improve service,
- facilities and collections are used effectively,
- the New Finnish Library Act will favor library as a meeting point in communities,
- staff-less service might be more convenient for customers who, e.g., work in shifts,
- the public will insist on staff-less libraries.

We will take a closer look on the reports on Finnish open libraries in the following Chapter 3.

Norway

The primary source of information in Norway has been the director of the Norwegian National Library (Nasjonalbiblioteket) Svein Arne Tinnesand (SAT) who has filled out and returned a questionnaire on August 8, 2016 (e-mail from SAT).

Basic Country Data

Norway has an area of 385,252 km^2 and a population of 5,213,985 inhabitants (2016). There are about 700–800 public libraries in Norway.

First Open Library

Concerning the year of the first open library in Norway, SAT tells that already in 2010, the central agency for archives, libraries, and museums (ABM in Norwegian) had granted money to a staff-less library project in Dyrøy community in Troms county in the Northern part of Norway. The project got great media attention, although it was never realized. Therefore, it became the Stavern branch library in Larvik, which is considered to be the first Norwegian staff-less library. The Stavern branch library opened in 2013 (https://no.wikipedia.org/wiki/Meråpent_bibliotek). Another source, the library director, Marit-Gro Berge (personal message, e-mail July 19, 2016) mentions another 2013 library branch, in Lindås, opened on September 6, 2013; this is probably the first Norwegian open library.

Present Number of Open Libraries

Concerning the present number of operating open libraries in Norway, SAT refers to the official library statistic from week 11 in August 2016, which tells that 71 out of 428 (16.9%) municipalities had staff-less main or branch libraries. The total number of staff-less hours per week is 1394, which corresponds to 11% of the regular, staffed opening hours. Concerning the Norwegian branch libraries, 25 out of the 428 municipalities have one or more staff-less branch libraries (in total there are 248 public library branch libraries in Norway). These more-opened branch libraries have 1420 staff-less opening hours per week, which corresponds to 34% of the 4166 staffed, weekly opening hours. In total, there are 78 staff-less libraries (both main libraries and branches) out of 678 units. It means that 11.5% of the public libraries in Norway were "more-opened" in August 2016.

Thus, the share of unstaffed opening hours is significantly different in main and in branch libraries. Compared to Denmark, the amount of Norwegian staff-less opening hours is much lower both in terms of the percentage of libraries with staff-less opening hours and the percentage of the staff-less hours out of the total opening hours.

Regarding the number of visits and loans at the Norwegian staff-less libraries, there are no central statistical sources, but data are sometimes collected by the individual libraries.

Data on user satisfaction are not collected systematically. However, SAT answers that many local reports have appeared in the media, indicating that the users very much appreciate the staff-less library model.

As to the location of the open libraries, SAT points out that they are found both in villages and in towns.

Concerning deliverers of library systems and technology to staff-less libraries, SAT informs that both BIBITS [Bibliotekernes IT-service (Information Technology Service of Libraries)] (a division within Bibliotekscentralen) (http://www.bibits.no/#!apent-bibliotek/c1tdt) and Biblioteksystemer http://www.bibsyst.no/produkter/apentbibl/ deliver staff-less library solutions.

Resistance

Resistance against the open library model has occurred in Norway; the leader, Monica Deildok, of the Norwegian Union of Librarians has, for example, expressed skepticism toward the concept in 2010 (http://bibforb.no/ubetjente-bibliotek/).

About who initiated the establishment of more-opened libraries in Norway, SAT points at the central library authorities as the primary and then both local library directors and local politicians as the secondary initiators.

Notable Libraries

SAT draws attention to the results of a recent project, funded in 2013 by the National Library, concerning the testing of different staff-less library models at three libraries (Spydeberg library (see box, Tjøme library and Vegårdshei library). The results of the projects are published in *Bibliotheca Nova*, the journal of the National Library, in 2014 (Ditmansen, Egeland, and Sama).

Vandalism

Concerning misuse and vandalism, SAT answers that there has been no reports in the media about vandalism. However, it is possible that some libraries have had such experiences and have not reported them to the media.

Trend

It is SAT's impression that the staff-less library model is growing in Norway. As "best practice" examples he points at the two town libraries in Tromsø and Sarpsborg and the above-mentioned libraries in Spydeberg, Tjøme og Vegårshei, which were supported by the National Library. Further information about Spydeberg library is found in a Box 3.3 in the following Chapter 3, besides a closer look on the reports on Norwegian open libraries.

Germany

The basic source besides library sites is Claudia Knauer, director of the libraries of the German Minority in Denmark (visited July 6, 2016).

Basic Country Data

Germany has about 82 million inhabitants and an area of 357,021 km^2.

First Open Library

The development of open libraries in Germany differs in more respects from the picture seen in the Scandinavian countries. First, the spread of the concept happened relatively late. Indeed, as late as December 11, 2014, opened the first modern open library in Germany. It was the branch library, Buecherhalle Finkenwerder, in Hamburg. The English term "open library" was used then, although it was also considered to find a proper German name for the new phenomenon. The Finkenwerder district is one of Hamburg's 32 library districts. The district holds shipyards and airplane factories but is also characterized by a rural atmosphere.

The Finkenwerder library was, according to the library's internet site, inspired by the positive experiences with open libraries in Denmark and the Netherlands. The present (September 2016) staff-less opening hours were chosen to be on Tuesdays and Thursdays from 1 to 2 p.m. and on Wednesdays and Saturdays from 8 a.m. to 4 p.m. According to the local plans, further opening hours will be added later in accordance with customer preferences. The open library was established after a successful pilot test. In connection with the opening, a number of changes were made, for example, new furniture and a clearer and more transparent structure (Wildermann, 2015).

German Minority Library in Denmark Was an Early Starter

It should also be mentioned here that the German public libraries of the German minority in the southern part of Jutland in Denmark in its branch libraries in several towns, already before the Finkenwerder open library, had established their own staff-less libraries, following a Scandinavian model (Wildermann, 2015; interview with Claudia Knauer, 2016). A similar but not identical self-service project was established in 2015 at a branch library, Neumarkt, in Cologne, with self-service each Monday from 10 a.m. to 6 p.m. However, because a security guard is present during the hours with self-service, the Cologne library case does not represent a genuine staff-less, open library model (http://www.stadt-koeln.de/leben-in-koeln/stadtbibliothek/news/stadtbibliothek...). The Finkenwerder open library branch holds the traditional facilities of a modern open library: RFID-based loan and delivery equipment, payment of fines and the like, Internet and library catalog access, and library card in-check. The public library system of Hamburg is Germany's largest with 4.8 million visitors per year and 460 staff members. It should, finally, also be mentioned that Claudia Knauer, the director of the libraries of the German minority in Denmark has been very active in spreading the open library idea to Germany.

Statistics

There are no available statistics but the text of the library site asserts that the customers appreciate the open library concept.

Vandalism

Although, vandalism has not occurred at the Finkenwerder branch library, Claudia Knauer emphasizes that the fear of unrest and vandalism is particularly widespread and common among German library professionals.

Sweden

The information on more-opened libraries in Sweden is provided by Kristina Elding from the municipality of Lund and complemented with data from digital developer,

Annelie Krell, from Kultur Skåne. Valuable information on the number of Swedish more-opened libraries are e-mailed from Cecilia Ranemo, researcher, responsible for the library statistics of Sweden, the Royal Library, Stockholm.

Basic Country Data

Sweden is a large 450,295 km^2 country with 9.8 million inhabitants.

First More-Opened Library

The first staff-less library in Sweden was opened in Veberöd in 2009 (Engström, 2015). The preferred term in Sweden is "more-opened" to emphasize that the staff has not been saved away.

Present Number of More-Opened Libraries

Kristina Elding points out that it is difficult to determine how many more-opened libraries there are in Sweden because of different definitions. Most of the more-opened libraries are found in the countryside in village branch libraries. There are, however, also some more-opened main libraries in smaller towns or in the central town of the municipality. More detailed information can be found in http://utveckling.skane.se/utvecklingsomraden/kulturutveckling/bibliotek/meroppna-bibliotek/.

According to the mentioned site there are 14 more-opened public libraries in the province of Skåne. Complementary data from Annelie Krell adds seven more library units, which have already opened or are planned to open in 2016. It gives a total of 21 more-opened libraries in Skåne. About the more-opened libraries in the other parts of Sweden, Cecilia Ranemo estimates that there are, totally, around 50 more-opened libraries in Sweden. The official library statistic, however, only counts the total number of more-opened hours for all public libraries in the individual municipalities and not the exact number of more-opened libraries. She also takes reservations as to certain public libraries, which report that they have only 1 h more-opened library per week. Such libraries are not what we, normally, consider as more-opened libraries (e-mail August 24, 2016).

Statistics

Here, Elding refers to the official library statistics of Sweden. She, however, points out that according to her own knowledge about more-opened libraries in Sweden, they had usually contributed to both increased loans and more visits.

The official Swedish library statistic delivers figures on the total number of weekly opening hours and the weekly more-opened opening in both main public libraries and branch libraries. Table 2.3 shows the latest figures from 2015.

Table 2.3 tells that 13% of the weekly opening hours in the Swedish branch libraries are more opened.

Table 2.3 **Staffed and Unstaffed Weekly Opening Hours (2015), Sweden**

Sweden	Staffed Weekly Opening Hours (2015)	Unstaffed Weekly Opening Hours (2015)
Main libraries	14,185	1221 (9%)
Branch libraries	18,697	2564 (13%)

http://biblioteksstatistik.blogg.kb.se/files/2016/05/Tabeller_Bibliotek_2015.xlsx.

Vandalism

Elding points out that vandalism and other types of negative behavior were feared by many, both professionals and users. However, the negative expectations have not been met. It is, according to Elding, one of the most important positive surprises after the more-opened libraries started, that they worked so well. She, here, emphasizes that the use of closed-circuit television in libraries are normally not found. The identification of the visiting user through their library card and PIN, coupled to the library system, is enough.

Resistance

The local press is usually positively interested in reporting on new more-opened libraries, especially in smaller municipalities. The users are often satisfied too. She also mentions that the Swedish library association has also showed interest and published debates on more-opened libraries (Hjalmarson, 2015a, 2015b). Even the Lund University has published a report on more-opened libraries (see further mention of Johansson et al. (2015) in Chapter 3).

Best Practice Examples

Here, Elding points out Kävlinge municipality as a place that has experimented with the more-opened concept. The municipality of Lund also has also experiences with more-opened libraries for a long time and there are several more-opened libraries in different districts of the municipality. In Jönköping county, there is an ongoing library development project with several municipalities involved and focusing on the physical library. http://plus.rjl.se/infopage.jsf?nodeId=43586.

Trends

More-opened libraries are certainly a rising trend. Many municipalities have recently begun to start more-opened libraries or they plan to do that. We will take a closer look on the reports on the Swedish open libraries in Chapter 3.

United Kingdom

The main sources of the section on the United Kingdom are VP Richard Stewart, responsible for Bibliotheca in the United Kingdom and Ireland, and Client Manager, Lisa Roberts, Peterborough City Council. Another information source was Kelly (2016).

Basic Country Data

There are more than 4000 public libraries in the United Kingdom. In 2015, they received 225 million physical visits. About half of the UK population has a library card. The public libraries are one of United Kingdom's most well used public services, the "community living room". The United Kingdom has 63,181,775 inhabitants (2011) and covers an area of 242,500 km^2.

First Open Library

The first open library in the United Kingdom was opened in 2013 in the Farsley library in Leeds city in 2013. There are no totally staff-less libraries in the United Kingdom.

Present Number of Open Libraries

There are (September 2016) 35 open library branches in the United Kingdom.

Vandalism

No cases of vandalism and misuse has been reported (Kelly, 2016).

United States

There is very limited information about open libraries in the United States. Wallin (2015) describes an open library in Washington state, Yacolt, from 2012.

It should be kept in mind that this version of the history of the public libraries represents an optimistic version. A more gloomy story would probably consider the development toward the open library as a further step toward the decline and fall of the public libraries in the Western world.

We will now take a look at relevant self-service experiences from outside the public library sector.

Public Sector and Business Contexts

Within the present public sector, at least in Denmark, we find a lot of various self-service models, using opportunities given by modern Information and Communication Technology. The applications cover a wide field from collection of taxes, social security, education to payment of fines, and notices of change of address. Usually, however, such functions are virtual and do not include access to the physical facilities of the responsible authorities or services.

Museums can also be characterized as self-service institutions because the visitor normally can view the collections without guidance. Normally, however, museums are totally closed outside the opening hours. Staff are also present as museum attendants, museum shop's staff, and the like. Therefore, a museum is not a staff-less institution. It is, certainly, so because the protection of the unique collections is of primary importance. And this critical task has not, yet, been taken over by machines and automatic devises. Of course, so-called "virtual museums" have appeared, but as they do not give direct access to the physical spaces of, for example, art galleries or archaeological collections, they are not comparable with staff-less libraries.

Should we point at a public institution with staff-less like opening hours, like many libraries, it could be protestant churches (at least in Denmark), where "Folkekirken" (the People's Church) is a public institution. Public parks, national parks, playgrounds, and the like are also among the relatively few examples of pure staff-less public services.

Private Sector Self-service Experiences

In business contexts self-service is a well-known concept, too, whereas staff-less is rarely used in the same way as in libraries. It is often claimed that the private sector, in general, is ahead of the public, in terms of innovation and service improvements. A closer look, however, often reveals that public sector institutions, at least public libraries, in terms of open access and staff-less units actually have been ahead of their private sector colleagues.

Indeed, a typical commercial self-service store is not staff-less at all. The manned cashiers, for example, are still an essential element in most supermarkets. One takes the goods from the shelves oneself but is then controlled and pays for it at the exit. Today, an alternative to the traditional cashier-staffed checkouts in retail stores exists.

Self-checkout or self-service checkout or semiattended customer-activated terminals are in operation in estimated half a million shops around the world (https://en.wikipedia.org/wiki/Self-checkout). The spread of self-checkout, however, has been moderate because the systems are highly vulnerable to shoplifting. In fact, shoplifters can easily fool the self-checkouts and the risk of theft is estimated to be five times higher than by the traditional cashier-staffed checkouts. The advantages are reduced labor costs, where one attendant can run four to six checkout lanes. However, attendants are still needed, and the customers are not allowed to unlock the doors of the retail shops. Therefore, even a self-checkout shop is not comparable to a staff-less library. In modern staff-less libraries, however, also the cashier's job, have, so to say, been automated.

Also concerning common self-service, the appearance of open access in libraries predates private sector self-service. It is often mentioned that self-serving stores were originally invented and even patented by Clarence Saunders in the United States in

1917. Public libraries in the United States and the United Kingdom with open access, on the other hand, have existed since 1879/1894. Black writes in this regard:

> *One of the most conspicuous developments in customer service before 1914 was the inauguration of the large department store, with its various titillating technologies like lifts, escalators, and roof-top gardens. However, one technology which the consumer patently did not experience, even in the modern department store, was self-service. This did not generally emerge until after World War II—except, that is, in the public library, by which time open-access self-service in lending departments had become the norm.*
>
> *Black (1994, p. 5)*

Self-service radically changed the basic relationships within the buying process. From a relationship, where the seller was put between the commodity and the customer to one where the commodity was put between the buyer and the seller.

In Denmark, the self-service concept was first implemented in 1947 in shops owned by the cooperative movement, which had started in both Great Britain and Denmark in the 19th century. However, the major breakthrough first happened after the economic boom in 1958 (Tapdrup Mortensen, 2011). Compared to the staff-less libraries there were both similarities and differences. On the one hand, self-service shops were hated by the small shopkeepers and loved by the customers. This situation is somewhat similar to what is seen—to a varying degree—in some libraries. On the other hand, an apparent difference is that although the long-term effects of the self-service shops were a tremendous concentration and closing down of small shops, staff-less libraries in many cases have contributed to the preservation of small branch units.

Self-service Technologies

It can be interesting and also relevant for library professionals to study the discussions and experiences achieved by the private sector because there are much more research concerning self-service models that stems from private sector enterprises than from public institutions.

Meuter et al. (2000, p. 50) point out that academic literature has focused almost exclusively on the "interpersonal dynamics of service encounters", although self-service technologies (SSTs) are increasingly changing the way customers interact with firms to create service outcomes. SSTs are defined as

> *technological interfaces that enable customers to produce a service independent of direct service employee involvement.*

Translated into a library context, open libraries are certainly an example of an SST application. Examples of business applications are Automated Teller Machines or cash machines (ATMs), automated hotel checkout, and, probably the most important, e-commerce and similar services over the Internet.

Such examples illustrate one important difference between the commercial self-service concept and the library staff-less model. The latter is bound to a physical library framework. However, the idea of vending machines in libraries has also been considered and recommended too.

Lending Machines in Libraries

Hampshire and Sanford (2009, p. 542) suggest

> *Increasingly, libraries are finding that machines based on vending technologies and the idea of self-service can be employed effectively to meet customer needs ... Just as people use self-checkout stations in grocery stores or libraries, or ATM machines and airport kiosks to perform routine tasks where they formerly needed the intervention of a staff member, the lending machine makes it possible to quickly and cost-effectively deploy library service out into the community into places where people gather.*

Different types of lending machines have been implemented in libraries in Canada, the United States, Sweden, and Finland since 2005 (Hampshire & Sanford, 2009, p. 543). The machines applied by the Contra Costa County Library in California were further modified to accommodate users in wheelchairs and also information in Braille was affixed. The machine holds up 400 paperback books, which a user can login, select a title, and have it dispensed in under a minute. People can borrow up to three books at a time and must return it in its case to the machine before the 3-week long loan period expires. The cost of each machine is about $100,000. The so-called "Library-a-Go-Go" brand was marketed intensely in California; the first machine was unveiled in 2008 (Hampshire & Sanford, 2009).

It is obvious that many of the benefits offered by a typical staff-less library are not present using at a "Library-a-Go-Go"—model: the selection of books, for example, is minimal with 400 titles and—most important—the access to the both inspiring and relaxing facilities of a well-equipped and well-furnished public library is also absent. Wallin (2015) finds book lending machines useful; but he also expresses reservations: book lending machines are expensive to purchase and they provide only a small amount of service. He argues, furthermore, that a local community for "similar start-up and operating costs, and with a little trust" in the community can provide an unstaffed library with a much more traditional feeling library environment, fewer barriers to use, and a lot more versatility (Wallin, 2015). The only point where the Go-Go-model, so to say, can surpass the staff-less, seems to be availability, measured in hours, where the Go-Go-model offers real 24/7 access."

Maybe, also, the purpose of the Go-Go-library is somewhat different. The location of the lending machines at busy railway stations and shopping centers indicates that a calm place for reflection had, apparently, not been the goal. It should also be mentioned here that similar tests in Sweden with Bokomaten had been stopped due to low use (Hampshire & Sanford, 2009, p. 547). Various, similar practices have taken place, in many countries, e.g., Germany (Öffentlicher Bücherschrank, 2016) since the 1990s. Often such free and take-away services have also been organized by others than the local public libraries, for example, artists, Lions-clubs, associations, and the like.

Customer Satisfaction With SST

Nevertheless, research results concerning private sector service encounters involving SSTs are relevant, when trying to identify sources of customer satisfaction and dissatisfaction. Another interesting research question relates to the issue to what extent the sources of customer satisfaction and dissatisfaction with SST encounters are similar to or different from the parallel sources with interpersonal encounters.

We will now consider the main results of Meuter et al.'s study (2000) on sources of satisfactory and dissatisfactory evaluations of different kinds of SSTs. The study was based on a qualitative methodology, the Critical Incident Technique (CIT), where a total of 1000 respondents have completed an online questionnaire. CIT is not meant to be generalizable, but rather to provide insight into the nature of the phenomenon studied (Meuter et al., 2000, p. 53).

A total of eight satisfying categories were identified:

1. Solved intensified need
2. Better than the alternative:
   a. Easy to use
   b. Avoid service personal
   c. Saved time
   d When I want
   e. Where I want
   f. Saved money
3. Did its job

Meuter et al. (2000), pp. 54–57.

We will begin by considering the relevance of those categories in a public library context. The first satisfying category, "solved intensified need", is exemplified by a person who has no money and who needs money for a taxi and therefore uses a relevant SST, a cash dispenser. Concerning the "Easy to use" category, we know that many users, apparently, find it easy to use the open library. We, however, do not know if certain user segments find access to and use of an open library more difficult than visiting the library during the staffed opening hours.

Avoid Service Personal

A rumor among the library staff in Denmark stated that certain user segments, for example, male parents, preferred the staff-less opening hours because they disliked to be disturbed by intrusive female staff members. However, such assumptions have not been either verified or falsified by research. So the status of the "avoid-service-personal" factor in library contexts is still undecided.

Saved Time

"Saved time" can certainly justify the use of a staff-less library. The most important is probably, however, the "when I want" argument. This justification is, particularly, important for busy career people and commuters. However, also unemployed person might have favorite times for visiting a library, which could lie outside the staffed opening hours.

Saved Time

The "where I want" justification is usually used in connection with access to services from off-site locations. Indeed, Meuter et al. (2000, p. 56) use an example with a library user who says: "I can checkout library books by phone without having to drive all the way down to the library to renew them." In other situations, however, the user could prefer the calm and relaxing environments of a public library, to a noisy apartment with screaming children. "Saved money" through knowledge-based action has always been among the key arguments for visiting a library, to get, for example, relevant consumer guidance.

Finally, there is the essential criteria: "you received what you needed". Here a wide spectrum of both material and immaterial needs in both commercial and library contexts could be relevant. Examples could be a needed book or piece of information, necessary relaxation to avoid stress, communication with other users, etc.

Meuter et al. (2000, pp. 56–57) also try to identify the dissatisfying aspects of SST, where

1. Technology failure
2. Process failure
3. Poor design
4. Technology design problem
5. Service design problem
6. Customer-driven failure

Technology failures have also appeared in staff-less libraries, especially, during the initiation phase. Process failures might also take place in an open library. Poor design can also appear. Indeed all dissatisfying aspects identified by Meuter et al. are relevant in library contexts too.

To sum up, the comparison between the specific library staff-less model and similar public and private sector examples illustrates the importance of distinguishing between a number of separate causes and elements.

One common element deals with the user's or the customer's free access to library materials and other goods. Although, many traditional public and research libraries and some grocer's shops, still contain counters the open shelves have become the norm in many situations. There are still contexts where free access is restricted, for example, concerning rare books and expensive goods, for example, diamond rings. As mentioned, it is probable that especially public libraries introduced "open shelves" long before open shelves became common in shops and commercial contexts. On the other hand, closed shelves were kept in some important research libraries even after 1970. As a student at the University of Aarhus, I personally experienced to use a research library with closed shelves.

The number of opening hours forms another dimension. In libraries, staff-less opening hours are often used to achieve prolonged opening hours, approaching the desired 24/7 standard. In supermarkets, for example, staff-less manning is seldom used only to prolong the opening hours.

Finally, self-service and staff-less service are different in one important respect. Self-service, typically, means that the customer selects his commodities and transports them—often by the use of special shopping trolleys—to the cashier, where a member of the staff will receive payment for the goods. Supermarkets, thus, offer self-service or sometimes also self-checkout, but they are, nevertheless, not staff-less as the staff-less libraries. In business contexts there surely exists staff-less solutions such as automatic teller machines outside banks, automatic gas stations, etc. Typically, however, such installations are found in the outdoors. It is extremely seldom that private business self-service solutions offer access to the homely surroundings, light and heat, and opportunities to stay for a longer time, features that exactly characterize the open libraries.

The open or staff-less library concept in different countries which was implemented a little more than 10 years ago, thus, both holds features inspired by business examples, but also some quite unique elements. Also the purposes and effects may show up to be quite different in business and in library context, for example, concerning the preservation of small branches. Besides, the open access revolution also shows public libraries in an apparent pioneer position.

To finish, I will put a small fly in the ointment, concerning the conception of libraries as the first on the staff-less service market. Actually, there exists one business where the customer locks into the facilities of the firm, and receives a service that can only take place indoors. Tanning salons resemble in many respects public libraries, apart from the less hedonistic and more intellectual and spiritual aspects related to library visits. Usually, there are no attendants present in the salons and the weekly opening hours are also extended. It is told that it was tanning salons, which, in particular, inspired the Danish open library pioneers, Mogens Larsen, to establish, the world's first running staff-less library in Gjern in 2004. However, one can argue that making a tanning salon staff-less is a more simple task than doing the same with a public library. Apart from a darker tan, the user does not take anything with her out of the salon. When she leaves a library with books and other library materials on loan, the technological challenges are greater. A staff-less tanning salon, for example, probably does not need RFID technology.

Summary

This chapter has focused on answering the two research questions:

- How has the staff-less library concept developed and how is this development to be seen in a broader library history context?
- How has the open library concept developed and spread in seven different countries in terms of primarily quantitative measures?

The idea of staff-less or open libraries appeared a little more than 10 years ago at two different corners of the world: in Denmark and in different countries in the Far

East (China, Japan, Singapore, and Taiwan). Although the environments were quite different, a rural thinly populated region of Denmark compared to densely populated urban areas in the Far East, the applied RFID technology was common in the two corners.

In a broader, library history context, the open libraries were the third of three historical opening revolutions (the first instances in parenthesis):

1. Open shelves (the open access revolution): the user has open access to the shelves, comparable to self-service in shops and supermarkets (the United States 1879/the United Kingdom 1894)
2. Automated self-delivery and self-checkout of library materials, unlike most supermarkets and shops with cashiers/checkout assistants (Australia 1991)
3. Open libraries (the open library revolution): the user can enter the unstaffed library's premises by the use of an identity card, unlike most supermarkets open libraries are unstaffed (Denmark 2004)

Concerning the spread of the open library concept in seven countries both the earliest and the strongest development took place in Denmark followed by the other Nordic countries, Finland, Norway, and Sweden. In contrast, the impact of the open library model in Germany is much more sporadic. In the United Kingdom, on the other hand, the open libraries seem to be in a take-off phase. The impact on public libraries in the United States is more difficult to measure.

The next chapter will focus on the development and history of open libraries, covering both the technological requirements and the development in different countries. Also variations as to the role of the main stakeholders in the different countries will be covered. Compared to this chapter, the treatment of the topic will be more based on qualitative research reports and articles.

The Development and History of Open Libraries

3

Introduction

The chapter starts with a presentation of the technological solutions behind the open library model. The section on technology owes a great deal to assistance and input from Bibliotheca and Cordura. Among others the section will also contain information on quite new trends allowing, for example, targeted communication with the users in the open library. The concept is called the "transformative library." Although there is, definitely, no "one size fits all" solution, there is nevertheless a number of key elements. Then a section will focus on recent reports and research on the Nordic open libraries and the open libraries outside the Nordic countries. Because the amount of literature on open libraries in different countries varies a lot, the treatment will mirror this variety. The same is valid concerning the focus and coverage of the different reports. Finally, there will be a review of the roles played by different stakeholders such as users, local politicians, and the staffs of the open libraries.

Technological Requirements

We will start by presenting an overview of the different technology elements behind open library solutions. Picture 3.1 shows the technical setup.

The system typically contains the following elements:

The Library Controller

The technical interface to the building.

- Door locks
- Light control
- Security alarm
- Sound announcements
- Public personal computers (PCs) and self-service stations
- Gates

The elements ensure that the staff has easy access to the control of the library and that existing equipment can be used with minimal adjustments. The all-in-a-box solutions offer simple relay interfaces that allow the staff to control their library manually or automatically. It is important that the staff procedures are kept simple and as close to "before open library" as possible.

Staff-Less Libraries. http://dx.doi.org/10.1016/B978-0-08-101923-8.00003-4

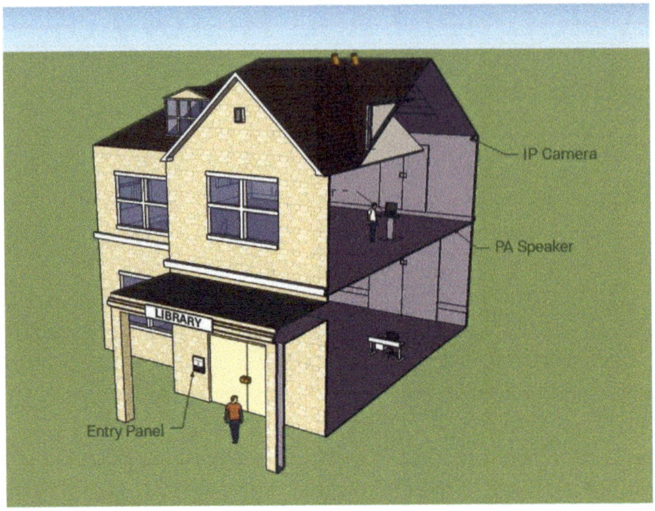

Picture 3.1 Open library elements.
Cordura (2016)

Staff should never be caught between suppliers like City IT and Security Alarm provider while unable to close the library.

Library Management System

User administration should always be kept in the library management system (LMS). Patron identity (ID) and pin code are in most places the key to the open library and can be retrieved from LMS by any of the many Application Programming Interface (API) standards like Standard Interchange Protocol (SIP2) and National Information Standards Organization Circulation Interchange Protocol (NCIP). Many libraries use patron categories and age to control who has access.

The open library solution should use a secure connection to the LMS and only store minimum of data outside the LMS.

The Entry Panel

The panel is the access control where the users let themselves in using their health card or library card. Some offers user-friendly panels with small screens that guide the user and offer different languages and information.

Video Surveillance

Typically, Internet Protocol cameras detect motion. This makes the review of events much easier for staff.

Public Address System (Sound Announcement)

A Public Address system delivers prerecorded messages to patrons when opening and closing. Many libraries also use announcements when the staff leaves and self-service hours begin.

The Administration System

Simple program that let the library staff change opening hours, see statistic reports, event logs, and video surveillance. There are different legal requirements in different countries, but in all cases data security is important.

Patron details shall always be encrypted and access to video limited to specific users and logged then accessed.

An open library is although being a physical library, very much focused upon a high-technological product based on an intelligent combination of advanced library automation and closed-circuit television (CCTV) technology. If we follow a user from the gate and till she or he leaves the open library we can identify the following stages or steps.

Open Library Technology as Experienced by the User

The first barrier as seen from the user's point of view is to unlock the door by the use of an ID card and a code/password. From both the users' and the library's point of view, it is important that the security identification works so that the proper users are not locked out and persons without ID cards, on the other hand, are not allowed to enter. The user's tolerance as to being rejected without a valid reason is probably limited. Moreover, the system should guarantee that only approved persons can enter. Most open libraries also emphasize that the entrance is video surveilled so that vandals and thieves can be detected, at least when they leave the library.

Then, the user has hopefully entered the library maybe accompanied by family, children, or friends. The technology should then ensure that library is enlightened and that the PCs, terminals, and other facilities also work. If the user, for example, wants to deliver borrowed materials the machines or kiosks should work, and the same, if she wants to borrow new materials and pick up reserved books. When the user wishes to leave the library, it is important that the systems detect if she takes with her materials that are not checked out properly. The systems are usually not able to see when a certain user leaves the library. It is also important that the systems can tell the user, for example, about when the library is about to close in the evenings and the like. It requires that audio-equipment is installed too.

Suppliers—Bibliotheca, Cordura, Axiell, and others

Apparently, the open library technology should handle complex interrelations between different systems (e.g., LMS, entrance) and functions that coordinate the entrance systems and lightning and audio systems with the LMS. This is a complex task that typically requires professional assistance.

A number of firms have developed specialized and innovative solutions targeting toward open libraries. Among them should be mentioned experienced actors, such as Axiell, Bibliotheca, and Cordura, who all have been engaged in establishing the earliest open libraries in the Nordic countries. In Denmark, Cordura A/S (system: Open Libraries) and Bibliotheca have been market leading since the world's first modern open library was established by Cordura in 2004. Later the high tech company, Cordura, has been engaged in the establishment of about 150 more in Denmark and in the other Nordic countries. Bibliotheca was also an early actor in Denmark in the municipality, Egedal, in 2009. Later about 250 open Bibliotheca libraries were established in the Nordic countries and about 50 in the rest of the world. Recently, in July 2016, the two companies have merged, which means that they now together have developed and maintain about 400 open libraries in Scandinavia and about 50 in the rest of the world, mostly however, in the European countries. Presently, the growth markets seem to lie in Finland, Sweden, and the United Kingdom. At the time, Bibliotheca has a central market position in the United Kingdom with its Open+ solution. However, the first open libraries have already also been delivered to Germany, Ireland, and the United States. Besides, France, Spain, and Switzerland have also showed interest in staff-less public libraries. Other companies operating on the open library management market are Axiell, where system Aperio was launched in April 2016 (source: e-mail to the author from the Danish director of Bibliotheca, July 1, 2016).

Technological Components

The open libraries are built upon both well-known technologies such as video surveillance cameras and equipment and access panels and more, innovative solutions. Especially RFID (radio frequency identification) is considered as an example of a revolutionary technology. Indeed, you cannot run an open library without RFID, but you can apply RFID without open libraries. The RFID is a very small chip that can be placed in the books or other library materials. The chip makes it possible to detect if the material with the mounted chip leaves the building or room without being checked out. Also, automated delivery and loan systems are required in a modern open library. However, as automated delivery and loan machines and self-service kiosks were already developed and installed in many public libraries all around the world already in the 1990s, the next step to the open library solution became easier. Here both bar codes and RFID-chips were applied that represents an example of an RFID application without an open library. However, bar codes in an open library would not work because they allow you to leave the building undetected. In manned libraries, the staff, of course, will be alert and keep an eye on the users.

A further division could be made between hardware and software components. Hardware could, for example, be door access control panels, the CCTV, the audio system, and the server, whereas the software facilitates that the LMS can "speak" to the radio-frequency ID on the door (Hitchcock, 2016).

Costs

Usually, the costs for establishing an open library can be divided into one-time capital costs and yearly costs. In many of the open libraries in the United Kingdom, the yearly costs are set at around £1500 a year, whereas the establishing costs vary a lot. Costs for supply and fit of the open library technology in the United Kingdom vary from £8000 to £15,000 and the preinstallation building costs were around £7000–9000. Besides, there can be substantial costs for renovating worn out premises, new furniture, and the like.

We will finally take a look at some recent technological innovation within the open library sphere.

New Trends/Opportunities in the Future

Companies like Bibliotheca and Cordura are constantly working on further development of the open library solutions so that they can continue to meet their customers or libraries wishes.

The recent Cordura developments are:

1. the first open library for the book mobile (the mobile library),
2. the remote locker, self-serviced, pick up and return of books,
3. "The Transformative Library."

The First Open Mobile Library

At the Danish Varde Library, the library bus has been transformed into an open library, just like a normal library. Alarm, door control, automatic light, voice alarm, and entry panel have all been installed. The bus is placed at a strategic good place for a week or two, and the users can enter the bus with their health card or library card, without any staff present.

The Remote Locker

The remote locker has been designed to extend the coverage and accessibility of the library service, offering patrons a self-service experience at locations and times more convenient to them. Deployable at any indoor location, remote locker allows the customers to reserve items in their usual way and collect them at the device, while also returning previously borrowed items.

The Transformative Library—A Greater User Experience When Visiting the Library

Imagine if the library could automatically adapt to the users' wants and needs… That is the reality with the solution "The Transformative Library."

The transformative library identifies the user groups that are in the library space and transforms the space targeting those at the library in that moment.

The user group is identified by collecting data about the users. For example, when they let themselves into the library during the staff-less hours, when using the self-service kiosks for delivery and loan of books, and when they move in certain zones in the library equipped with motion sensors.

These data are processed and the results are used to digitally activate different "modes" at the library. "Modes" should be understood as a theme or a topic, chosen from the above input. The different modes can launch music playback, advertising on speakers, lighting changes, and presentations on large screens such as recommendations from the library, movie trailers, etc.

Using The Transformative Library, the local libraries have the opportunity to offer a more focused and thus better and clearer communication with the users—not at least at the self-managed hours.

Targeted Communication at the Danish Open Libraries in Randers and Silkeborg

At Randers Library South and Them (around Silkeborg) Library, the solution is already running. Here, data on sex, age, preferences of genres, and current interest are collected. Based on the data, the system calculates which mode should be activated. For example, if the data indicate that families with children are the dominant group in the library right now, the mediation on the screens changes, so that relevant info for this particular group will show up, music will be played in the children's department, and speakers will advertise library and other events relevant for families.

At the two libraries eight different modes can be activated:

- Children 0–6 years old,
- Children 6–12 years old,
- News,
- Crime,
- Journals,
- Fiction,
- Movies, and
- Travel/Food.

Each mode consists of targeted communication that can provide the user with inspiration and present new opportunities.

A Scalable Solution

The project aims, among other things, have been to achieve a better user experience for the library users, by using digital transformation of the library space based on factual knowledge of the users. By using real-time user presence for targeted communication it represents a brand new initiative within the library community. The system is fully scalable in relation to the data that identifies the user groups, as well as the changes that take place. It is actually only your imagination (and the technique) that sets the limits.

At the two open libraries in Randers and Silkeborg, there are many staff-less opening hours each week, but the solution can also be adapted to a large main library.

The project is developed with the financial support of the Danish Agency for Culture and Development for public and school libraries.

Further Development of "The Transformative Library" in 2016

In 2016, the concept will be further developed with the new project "Transformation of New Library Types."

The project is developed by the Danish libraries in Randers, Silkeborg, Viborg, and Cordura with support from the Development Pool for public libraries and educational learning centers under the department of the Culture Agency.

For more information, visit the project Website at http://www.transformativtbibliotek.dk and Facebook page https://www.facebook.com/transformativtbibliotek/.

Links to articles about The Tranformative Library: http://www.infotoday.eu/Articles/Editorial/Featured-Articles/Using-real-time-data-to-transform-a-library-112360.aspx?utm_medium=email&utm_source=flipboard.

We will now in the next two sections take a look at some of the reports and articles that have been written on open libraries in the Nordic countries and in the rest of the world.

Open Libraries in Seven Countries Based on Research Oriented Reports and Articles

This section contains a review of reports and articles—mostly in peer reviewed journals—on open libraries in the four Nordic countries: Denmark, Finland, Norway, and Sweden. Some aspects of the mentioned report may, already, have been referred to and it will also occur that some will be mentioned later in specific contexts. The index will contain the references.

Denmark

The Danish open libraries have been diligently referred to and commented, more or less critically, in newspapers, local newspapers, and different journals. Also, the Danish national library authority under different names (Styrelsen for biblioteker and medier and later: Slots-og kulturstyrelsen) and the Danish Library Association have given attention to the topic in reports. Finally, three articles have been or will be published about open libraries in peer-reviewed journals. The first report commissioned by the national library authority, the Danish Agency for libraries and media [Styrelsen for biblioteker og medier] was published in 2012 (Johannsen, 2012a).

Open Libraries—A Mapping Project (2012)

The report was authored by Carl Gustav Johannsen (2012a). It was based upon a quantitative questionnaire survey, qualitative interviews, and observations (visits to selected open libraries). An important source was data on visits and loans that 34 open libraries made available by the library software firm, Cordura. The data for the survey were collected in October 2011 by the agency with an impressive response rate of 90%. Eighty-one open libraries returned the questionnaires.

The more and less filled out questionnaires allowed to answer a number of important questions about the emerging library type. About the different backgrounds for establishing an open library, saving and budget cuts were believed to be more common than the answers indicated. However, the survey stated that budget cuts were a primary cause in about one-third of the cases. As to the neighborhoods chosen for open libraries, only 7% of the open libraries at that time were found in urban settings.

Visits and Loans

Another finding of the survey about the usage (visits and loans) caused a small sensation because the data showed that visits and loans per hour in all the 81 cases were significantly higher in the staffed than in the unstaffed opening hours. Open library critics paid much attention to that particular finding (Lerche, 2012a, 2012b). The survey results also showed that the establishment of an open library sometimes but not always was accompanied by increased numbers of loans.

The Users

The survey only contributed a little to show who the open library users were and how they probably differed from the users of the library during the staffed opening hours. The Cordura figures, however, could only tell about the age and gender of the open library users. Later surveys such as Moos-Bjerre and Moos-Bjerre (2014) and Broe Sørensen (2013) have been a genuine quantum leap in this respect by adding much more interesting details.

Based on observations, interviews, and miscellaneous sources a number of critical success factors (CSF) concerning open libraries were identified. The CSFs are referred to more detailed in Chapter 5.

Staffless Libraries—Recent Danish Public Library Experiences (2012)

Johannsen (2012b) reports on Danish experiences with staff-less public libraries in terms of local community characteristics, their usage in terms of visits and loans, characteristics of their users in terms of sex, age, and, finally, an analysis of critical success factors revealed. The article is to a large extent a summary in English of the Danish report *Open Libraries—A Mapping Project* (2012).

The Libraries of the Future—Target Group Oriented Knowledge for Library Development (2014)

The report (Moos-Bjerre & Moos-Bjerre, 2014) will be much referred to later. Therefore, the presentation here will become rather brief. The report is not a report devoted entirely to staff-less libraries—the scope is much broader. However, it is the first representative survey documenting that the library users in Denmark appreciate the open libraries very much. The report is based on online questionnaires to 2000 library users and nonuser. The *Libraries of the Future* survey was ordered by a Think Tank, set up by the Danish Library Association.

Better Libraries 2013—National Results (2013)

Like the report mentioned earlier, this report (Broe Sørensen, 2013) is not devoted only to staff-less libraries. Like Moos-Bjerre and Moos-Bjerre (2014) it is a representative quantitative survey. However, the data are collected differently. Because of that the *Better Library* figures allow comparisons between municipalities that produce interesting insights. A total of 27,182 users of 179 libraries in 42 municipalities have contributed to the 2013 report.

Personae

The personae created are somewhat different from the personae of the *Libraries of the Future* survey. Four types were developed:

1. The big user
2. The dedicated user
3. The student
4. The borrower

Open Libraries Means More Visits

The report's main conclusion concerning open libraries is

> *a general and distinctive perception that open library means that you visit the library more*

This conclusion is derived from the level of agreement with the statement: "The open library has led me to use the library more." Three percent "strongly disagreed" and seven percent "disagreed." A total of 20% took up a neutral position. A total of 27% and 37% either "agreed" or "strongly agreed," which is a remarkable result. Open libraries are not only appreciated by the users but they also stimulate and increase the use of libraries.

Discrepancy Between Open Library Users and Nonusers

The *Better Libraries* report also reveals an attitudinal discrepancy between open library users and nonusers. As soon as a nonuser has visited an open library herself, the data show, that her attitudes often change in a more open-library friendly direction. This finding is supported by a revealed pattern showing that user attitudes toward open libraries are at least changed in municipalities with only one or very new established open libraries.

A total of 40% of the respondents had visited an open library. A remarkable detail revealed is that although the library used in the staffed and in the unstaffed opening hours is typically the same unit in smaller municipalities, in larger communities the staffed and unstaffed opening hours are more likely to be spent in different units.

User Attitudes and Satisfaction

Another interesting pattern revealed by the 2013 *Better Libraries* report concerns user attitudes toward the staff. Swedish data indicate that the user's enthusiasm for open libraries do not require a disapproval or dissatisfaction with the staff. Although the 2013 survey reveals that the staff is more seldom used than in 2011, the users' satisfaction with their staffs is high, 90 on a scale from 1 to 100. The users are apparently still wild about their librarians; 94% either "agree" or "strongly agree" with the statement that they get the help they need and 91% "agree" or "strongly agree" that the staff is visible and attentive to the needs of the users. Service and the experience of the library as a nice place to be here count the most. In the 2013 *Better Libraries* survey, opening hours count less. It is interesting compared to the 2014 *Libraries of the Future* survey, where opening hours were among the highest ranking library interests.

The 2013 survey also reveals interesting details about the purposes the staff are used to serve. Information seeking assistance and technical problems have apparently become less important tasks compared to finding materials on the shelves, inspiration, and practical help.

The survey also highlights important aspects of the users' perception of the open libraries; 19% "agree" and 12% "strongly agree" with the statement that the interior design and the arrangement of materials in the open library inspire me to new loans. This result, certainly, confirms that it makes sense and pay off for libraries to focus on design and user-friendly arrangements. Even more successful are the libraries to have been in creating orderly arrangements. Indeed, 49% "agree" and 57% "strongly agree" with the claim that arrangements in the open libraries are orderly. Finally, do 92% either "agree" or "strongly agree" with the statement that the opening hours fit me.

Barriers

The survey also addresses barriers associated with the open libraries. As to technical problems, a vast majority, 91% of the open library users either "agree" or "strongly agree" with the statement that it is technically easy to use the open library. Among the nonusers the percentage was 44% of the open library users who feel safe in the open library, whereas 43% of the nonusers did. A total of 77% of the nonusers preferred staffed libraries, whereas 42% of the open library users did the same. Finally, there are also 42% who indicated that they did not care if there were staff members present or not. Maybe it should be mentioned that 35% indicated that they adopted a neutral position on this particular issue.

Innovative Public Library Services—Staff-Less or Staff-Intensive (2014a)

Johannsen (2014a) here discusses whether recent library innovations such as automated loan and delivery systems, virtual services, and staff-less libraries open 80 h a week will make professional and clerical staff superfluous. Based on empirical data from two 2011–12 Danish library development projects, Johannsen concludes that staff-less and staff-intensive public library service concepts can both coexist and also create advantageous synergy in the library. The study, however, also reveals that staff attitudes toward both staff-less and, more surprising, also toward more staff-intensive practices have been somewhat skeptical and reluctant. The article also presents leadership initiatives, which have been tested successfully to handle such resistances.

Strategy as Legitimacy Work—Strategic Answers to the Challenges of the Libraries (2016)

Kann-Rasmussen and Tank's (2016) journal article compares, through the use of discursive policy analysis, three recent Danish public library development strategies. The open library strategy is one of the three strategies. The analysis shows that availability and effectiveness are common elements focused in all three strategies. The analysis also shows that this particular focus is implemented at the expense of other concerns such as library professionalism that the article finds problematic.

Finally, I would mention a Ph.D. project, hosted by Copenhagen University on staff-less libraries in Denmark and Sweden. The Ph.D. student, Lisa Engström, reports in Box 3.1 on her ongoing project.

Box 3.1 An Ongoing Research Project on Open Libraries

Libraries minus librarians, equals what?

A comparative user-oriented study of staff-less public libraries in Denmark and Sweden.

Lisa Engström, Ph.D. student, Royal School of Library and Information Science, Copenhagen University.

Today, more and more public libraries in more and more countries implement staff-less opening hours. In Denmark, the first staff-less library opened in 2004 and in Sweden the first opened in 2009. Librarians are traditionally in the heart of our understanding of libraries. Therefore, this development is striking and raises several question such as: How does users perceive staff-less libraries? Are users experiencing improved accessibility when they are able to access the library at times which they prior to the implementation of staff-less opening hours had not been able to do so? Are users experiencing problems when navigating the library by themselves? If so, how do they perceive that? Moreover, what are the users actually doing in the staff-less library?

Given the increasing number of staff-less libraries, the questions on how staff-less libraries affect the library institution and the users' perception of it are vital. In my Ph.D. project, I explore staff-less libraries from a user perspective. In-depth interviews, structured observations, and text analysis of policy documents allow me to analyze users' experiences and perceptions, as well as the users' actual practices in the library and the social and political context. Indeed, I want to explore and analyze how users experience the library when they are navigating it themselves and I want to examine what they are doing in the staff-less library at a daily basis. The research takes place in Denmark and Sweden at public libraries with staff-less opening hours.

The context in which staff-less libraries are situated is versatile. Technological changes as well as social and political shifts affect how staff-less opening hours are implemented and how the users perceive it. To apprehend the phenomenon of staff-less libraries more fully, I emphasize the significance in considering this context. Therefore, when analyzing the staff-less library, I will relate to a broader picture and discuss how the users' experiences and practices intertwines with a changing context, including for example what is usually referred to as a library crises and an economization of the cultural policy.

By Lisa Engström.

See also the recent article (Engström & Eckerdal, 2016).

Libraries Minus Librarians, Equals What? A Comparative User-Oriented Study of Staff-Less Public Libraries in Denmark and Sweden (2016–18)

Finland

In Finland, staff-less libraries have been the topic of a few academic theses. In 2015, Mika Mustikkamäki published a survey about the self-service libraries. It was published by Regional State Administrative Agencies both in Finnish and Swedish: *Undersökning om meröppet på de allmänna biblioteken i Finland* [*Survey of more-opened in public libraries in Finland*]. The data for the survey were collected in 2015; since that the development has been very rapid. Some of the statistics are, therefore, already now out of date.

There are also some examples of theses, prepared in the universities of applied sciences, dealing with self-service libraries. The theses are all written in the difficult Finnish language; however, their abstracts are in English (Blomqvist, 2015; Ollikainen, 2015). These theses will, however, not be reviewed here.

Survey of More-Opened in Public Libraries in Finland (2015)

Mustikkamäki's 2015 report presents the result of an online questionnaire survey sent to municipality public library directors in Finland in week number 8 in 2015. The mailed questionnaires were answered by 221 respondents, representing a response rate of 76%. The basic, filtering question was whether the library had already implemented, or planned, in 2015 and 2016, to offer unstaffed library premises to its customers. The choice of the word "premises" is interesting because it allows a library with, for example, only a staff-less newspaper reading room to count as an open library. The first staff-less library in Finland, which opened in 2009, was actually "only" such a staff-less newspaper reading room.

Sixty-one of the participating library director respondents confirmed that they were offering staff-less library facilities. The survey, however, does not tell how many weekly openly hours we talk about. Is it a 24/7 model or something less? Concerning the weekly opening hours one question is able to shed some light on the problem. Answers to the question about opening during weekends (Mustikkamäki, 2015, Figure 17) show that 48 units are open during weekends, whereas 13 are not.

The applied terminology is interesting too. In Finland, the Swedish term "meröppet" or in English "more opened" and "increased openings hours" are the two terms now, typically, used in connection with staff-less library issues. In Finnish, a language belonging to the Finno-Ugric language family, open libraries are called "amatoimi kirjastat." The term self-service, on the other hand, in Finland now, mostly, seems to refer to what the users do themselves, e.g., automatic delivery and loans, during the staffed opening hours.

Fifteen out of the 61 indicated that they had received financial support by state agencies to develop their new staff-less library services. It is also interesting to take a closer look at the reasons why the 61 directors have decided to establish staff-less library services.

Here, 19, about a third, pointed at savings as a purpose (Mustikkamäki, 2015, Figure 5). For comparison, 54 respondents mentioned increased use of their facilities, 57 increased opening hours, and 56 improved services as staff-less library goals. The third mentioning savings as a goal, corresponds with similar figures from Danish surveys. A further question regarded savings. Here, the respondents were asked whether the staff-less opening hours had led to fewer FTEs (Full Time Equivalents) and other service level cuts. Nine out of 61 answered "yes" and 52 "no" (Mustikkamäki, 2015, Figure 19). These figures seem a little bit inconsistent with Figure 5, where 19 pointed at savings as a purpose. Nevertheless, the data confirm that savings were not experienced as a primary purpose of the open library development by Finnish library directors in 2015.

Age Restrictions

In the debates in Finland about staff-less libraries the restrictions as to age had been mentioned. Critiques have pointed out that youngsters do not often have access alone to the open library premises. Here, 25 indicated that there was access for all, whereas 12 directors had a limit from 18 years and 8 a 15 years limit (Mustikkamäki, 2015, p. 11, Figure 6). Certainly, children accompanied by adults will have access to the libraries, but will usually not be able to lock themselves.

Camera Surveillance

Camera surveillance occurs in most of the libraries. However, 8 indicates that they are not monitored by CCTV. In a quarter of the instances, a watchman company had the supervision of the library during the staff-less hours. The questionnaire also included a question about the nature of the card required for access to the library. Here, 48 applied cards were usable for more purposes, whereas only 13 used specific cards targeted at the single library unit or municipality.

Vandalism

An interesting question regarded the occurrence of vandalism and similar problems associated with the staff-less libraries. A significant majority, 57 out of 61, indicated that they had not experienced any disturbances of the public order associated with the staff-less libraries. Here, the Finnish experiences with staff-less libraries very much correspond with the Danish.

Material Restrictions

The survey also included interesting questions regarding the library services provided during the staff-less opening hours. Here, the questions whether all library materials were available for loan during the unstaffed hours. Here, more than half of the respondents answered "yes," whereas the rest, 26, said "no." The restrictions were, primarily, associated with music and film media. Some library materials are associated with an age limit. It is interesting, here, to find out whether such age restrictions are respected during the staff-less opening hours. According to 49 out of 61, the library system was able to control that youngsters were not supplied with unsuitable for their age library materials.

User Satisfaction

Finally, the respondents were asked a number of questions about customer satisfaction measurements and library usage monitoring. Only about a fourth, 13 out of 61, of the respondents have measured customer satisfaction as to the staff-less library service. Most of the respondents, however, measure the number of visits.

The qualitative evaluations range from excited statements such as

- "Staff-less libraries support a user-oriented library development"
- "Staff-less libraries enable more qualified time for the professional staff"
- "Staff-less libraries facilitate a transition from routine barcode reading to professional library work"
- "Staff-less libraries support the use of library premises for meetings"

Box 3.2 above reports about Finnish experiences with open libraries around the capital, Helsinki. It is interesting, for example, to learn that the percentage of users who visits the library during the unstaffed hours within 6 months has risen from 15% to 25% and at the moment is around a third of the users. Also, the observation that the most popular time to visit an open Point library is Sundays between 12 a.m. and 1 p.m.

More negative critical opinions are

- "Staff-less libraries are contrary to the library law by restricting youngster access"
- "Staff-less libraries restrict access for visually impaired, persons with reduced mobility, and mentally disabled users"
- "Staff-less libraries make personal service disappear" (all quotes translated by CGJ)

In Box 3.2 the director of the Vantaa library in Finland, Mikko Vainio, reports on the increasing use of the library during the unstaffed opening hours. It is interesting that he tells that Sundays from 12 a.m. to 1 p.m. are the most popular time to visit the library. It is also remarkably that the unstaffed opening hours, primarily, are used for other purposes than borrowing books. It is also worth mentioning that the segment "well-educated, busy young families" is emphasized as the core users of the Finnish Vantaa.

Norway

Because there apparently are no Norwegian open library reports, their experiences are here represented by an account from Spydeberg where the open library model was introduced in 2014. One interesting report, *Brukeradfærd på storbybibliotekerne*, 2015, could have been mentioned here. Instead, however, it will be commented on in Chapter 4.

Sweden

A few comprehensive surveys have been published until now. Among the most interesting is the 2015 survey, *Meröppna bibliotek – en verksamhetsform för alla?* (in English: *Staff-less libraries – a concept for all?*) by G. Johansson, E. Lindberg, and J.R. Eckerdal, all researchers from the Department of Cultural Studies at the University of Lund (2015). Another survey contains the evaluation of one, specific Swedish open library in Veberöd

Box 3.2 The City of Vantaa and Its Libraries (Finland)—Case

Vantaa is the fourth largest city in Finland and growing rapidly. It is part of Helsinki Metropolitan area, where one-fourth of the population of Finland lives. Ten Vantaa libraries serve annually two million visitors and give out three million loans.

The PIAAC (Program for the International Assessment of Adult Competencies) survey 2012 confirmed what we're daily witnessing in libraries: there are among us many who lack basic skills of literacy, numeracy or digital skills (OECD, 2015).

Libraries are in business of advancing literacy of all. In order to succeed we have to compete of time of our users. Generous opening hours are a key factor, if only the cost of labor were not that high!

Open Libraries, open issues.

Since August 2013 we've been experimenting with Open Library technology. The first library was Point Library (Actually, Point Library is an acronym from "Pakkalan Oppimis- & Informaatiotalo" Learning and information house in the suburb of Pakkala). The library shares the building with The International School of Vantaa and English kindergarten, hence the English-sounding name. In Finnish we call the library Pointin kirjasto, situated in a building with almost every component already there: electric locks, lighting, cameras, etc. Also that particular neighborhood was an ideal place to start: well-educated, busy young families, who'd find this kind of concept exciting, promising, and useful.

It was not whether we would like to have one but how to start operation? There were no previous installations in Finland of this scale. Among unanswered questions were for example: who should be able to come in after manned hours? When the staff leaves, should all current users register themselves at the door? How about children? How should the premises be locked and emptied of anything of interest to thieves? What to do in case of emergency? How would people behave?

The library management was very supportive and sent the local staff to visit libraries abroad. Point employees got free hands and the power to anytime cancel everything if they so felt. Based on that, they dared to make the Open Library as open as possible, and later tighten the rules, if needed.

Since then, we have allowed all Helsinki Region library card-holders to enter the library, regardless of their age or where they live. The potential user base is over 800,000 large.

Point library has been open every day since January 2014. The only exception is the New Year's Eve, when rockets are fired all around. Also, when there's three or more days of public holiday in a row, Point library can't be open all the time due to accumulating returns.

Unmanned opening hours have steadily gained popularity. During the first 4 months of operation, 15% of visitors attended during the unstaffed hours. After

Continued

Box 3.2 The City of Vantaa and Its Libraries (Finland)—Case—cont'd

6 months it was 25% and at the moment one third uses the library during the unmanned time. Actually, the most popular time to visit Point library is Sundays between 12 a.m.–1 p.m. when there's no staff present.

The amount of materials borrowed has risen only slightly, as most visitors seem to work or study or read magazines when visiting the library.

Salaries form the single biggest cost of operation in our libraries, about 60% of the total running costs. In the case of Point library the costs are a little lower now, as users learned to be more independent. There are no other extra running costs, as the system is actually quite simple.

By Director Mikko Vainio, Vantaa Library, Finland.

Box 3.3 Spydeberg Public Library (Norway)—Case

In 2014, Spydeberg public library opened the doors to a new staff-less, open solution, 6 a.m. to 10 p.m. every day. Spydeberg is a small municipality with 5600 inhabitants. The library staff is currently 2.1 full time positions. Before the opening the library didn't have a self-service kiosk. Patrons waited in line to check out materials at the staffed counter. Often they didn't bring their library card, since the staff knew their names. After the opening we put a lot of effort into showing the new equipment and encouraging patrons to check materials in and out by themselves.

The Room

The new machines were not the only innovation. We painted, put up new signs, and rearranged the books and films to make it easier for the patrons. We hired an interior designer who helped us to create a brighter room with shelves following the line of light, and no tall shelves on the floor. She argued that in an open, staff-less library patrons should feel safe, and to do so there can't be any seemingly scary corners. We liked the new, open space, but we also learned that the patrons wanted to sit out of sight. We solved this by buying sofas with a high back for the magazine section and we placed a new, comfortable arm chair in a corner in the new crime and suspense section.

The Way We Work

We have a small work station close to the entrance, allowing us to greet the patrons as they enter the main room. Self-service has freed time, but we still have to work with book loans and getting materials ready while at the work station. We also have an x box for the young to play in the library and keep the games at the work station. Ideally, we would walk the room with a laptop, but this isn't an option for us. Nevertheless, we are more accessible than before and have time for

Box 3.3 Spydeberg Public Library (Norway)—Case—cont'd

a chat with those who want to, and better time to help those who need it. A few of the older patrons won't use the self service, and we help them as best as we can.

What We Learned

The age limit for accessing the library during the staff-less hours is 18 years. At the beginning we granted everyone over 18 admission without any further ado, and when staff hours were over we said goodbye to the patrons and left. A risk assessment put an end to this. Now patrons sign a contract and we tell them about fire escapes and safety measures. When the staffed hours are over, we empty the room. Those with admission then let themselves in again. This way we know who has been present. We have had a few incidents in staff-less time, which we have addressed. We had to rethink our policy regarding access to the open library and have had to exclude some patrons for a period of time, but in sum we feel that we offer the patrons something unique that they have received well.

What Patrons Think

It's a success. Patrons enter at all times, and express their gratitude at the trust they are given. The open solution has made it easier for them. They also appreciate and use staffed hours, as the staff-less hours are a supplement to the traditional library service, not the other way around.

The Future?

Spydeberg is facing budget cuts and we see that the open, staff-less solution is being used as an argument to cut down on personnel. If this happens, we will have to cut down on staffed opening hours as well. This was never the intention, we meant to develop a better library service and so far, we have.

By Ibi Engsby, library director.

that was the first, Swedish open library. The title was translated into English: "*Thanks the more opened library there exists a library even for me" -an evaluation of the more opened library in Veberöd 2010–11*" (Skåne, 2011). This report will be reviewed after the 2015 report. Finally, two articles from 2015 and 2016 (Engström, 2015; Engström & Eckerdal, 2016) based upon an academic master thesis will be dealt with.

Staff-Less Libraries—A Concept for All? (2015)

The survey covers the experiences gained at nine staff-less libraries in Skåne in the Southern part of Sweden, closest to Denmark. Like in Finland, the preferred term in Sweden is "more opened" (meröppent). The nine libraries are all situated in rural environments in the villages Veberöd, Glumslöv, Kävlinge, Löddeköpinge, Strövelstorp, Hjärnarp, Jonstorp, Anderslöv, and Rydebäck. The data were collected in the autumn 2014 through a diversity of qualitative methods, primarily focusing on local community consequences and effects on the professional work tasks and workload in the libraries.

The four principal research questions were

1. Are there any common traits associated with a librarians work in a "more opened" library and what traits should be mentioned?
2. Are there any characteristics of a "more opened library" in terms of demographic variables, geography (village/town), and opening hours and which are these characteristics?
3. How are the "more opened" libraries used?
4. Which strategies lie behind the introduction of "more opened" activities and to what extent have they been changed?

Most data stem from 23 librarians and library directors from seven of the nine libraries. However, also a few users were interviewed. Web-based questionnaires with open questions were combined with four in-depth interviews. The survey is especially interesting because it tries to look upon the staff-less libraries in a broader, theoretical cultural policy, sociological and psychological perspective. Theoretical frameworks by, among others, Oldenburg (1989), Goffman (1990), and different relevant library-oriented theories, e.g., on serendipity (Björneborn, 2008), library anxiety (Eklof, 2013), social capital (Vårheim, 2014), and urban development (Hvenegaard Rasmussen, Jochumsen, & Skot-Hansen, 2011; Jochumsen & Hvenegaard Rasmussen, 2000) are presented to achieve a better understanding of the open library issue.

Why Have More Opened Libraries Become an Important Issue?

The survey considers why staff-less libraries have become an important issue on the cultural policy agenda today. Here, the report points at both a recent trend, the 24/7 society, and at the technological developments, which have made the open library service possible. Here, the technological development is emphasized as being extremely important. The considerations about the relationships between the Internet evolution and the staff-less libraries are also interesting. It is assumed that the Internet had created new expectations as to the availability round-the-clock. These increased expectations have then been transferred to the physical library too.

Comparisons Between Sweden and Denmark

The report also contains remarkable comparisons between staff-less library development in Denmark and in Sweden. It is emphasized that the Swedish staff-less library development, unlike the Danish, is primarily triggered by a wish for service quality improvements to adjust to changed user preferences, whereas the Danish is motivated by a desire for budget cuts. It is true that about one-third of the Danish staff-less library establishments in the autumn 2011 had been influenced by budget cut considerations (Johannsen, 2012a, 2012b). However, as it has been pointed out earlier, sometimes the question in some Danish libraries seems more to concern different priorities (local community protection versus strong library units) than pure budget cuts. Furthermore, the Swedish more-opened library critiques have claimed that budget cuts also constitute the underlying agenda for the establishment of staff-less libraries in Sweden (Hjalmarson, 2015a, 2015b).

The Veberöd 2009 Story

The story about the background for establishing the first Swedish staff-less library in Veberöd in Skåne in November 2009 is also told. The initiative came from the city library in Lund where a survey had been done in connection with the planning of a new library if Lund should succeed in becoming European cultural capital in Europe. However, Lund was, unfortunately not, chosen as a European cultural capital and the planned new library was therefore not established. However, a user survey had been done that indicated that longer and increased opening hours were among the most desired wishes of the citizens. Two librarians at the city library in Lund then decided, after having studied Danish staff-less libraries, in concert with the city librarian, that Veberöd had the right background for testing the staff-less concept in Sweden (Johansson et al., 2015, p. 2). This story clearly shows the importance of documented and measured user preferences. A recent Danish survey (Moos-Bjerre & Moos-Bjerre, 2014) reveals a similar picture of users preferring flexible and extended opening hours.

Staff Attitudes Before and After

Another interesting topic, covered by the Swedish report, is the attitudes and feelings of the staff both before and after the introduction of more-opened library services. Here, data from the survey indicate that the feelings and attitudes considerably changed in a more positive direction after the implementation of the staff-less model. After the implementation, 56.5% of the respondents expressed very positive and 43.5% positive feelings toward the new more-opened model. Besides, a clear majority, 93.5% (22 out of 23) confirms that they understand the reasons why the more-opened library model had been implemented (Johansson et al., 2015, p. 13). A comment goes

> *People are used to do one's shopping later or you use the net to order articles and food ... I believe that you more and more demand a company which is open at odd hours.*
>
> *Johansson, Lindberg, and Eckerdal (2015, p. 14).*

It should also be mentioned, here, that even before the staff-less model was implemented, only 4.3% had very negative feelings toward it. A characteristic staff reply sounds

> *I've initiated "more opened" together with the library director and I consider it to be a natural development of the library service*
>
> *Johansson, Lindberg, and Eckerdal (2015, p. 14)—translated by CGJ.*

We will now take a look at some of the theoretical frameworks used in the report to better understand the more-opened library concept.

Third Place—Oldenburg

The "third place" concept was coined by the American urban sociologist, Ray Oldenburg (1932–). According to Oldenburg the home and work make up the first and the second place. Third place was used by Oldenburg in his 1989 book, *The Great Good Places*, to describe places for social interaction, apart from home and work place. "Third places" are informal public gathering places. In his book, Oldenburg argues that third places are valuable for both the democracy, the civil society, and for people's active participation in both. It is remarkable that libraries are not included in Oldenburg's enumeration of "third places"—*cafés, coffee shops, community centers, beauty parlors, general stores, bars, hangouts"*—in the subtitle of his famous 1989 book.

The Swedish report agrees that "more opened" libraries are able to contribute to the development of third places in the society. It is also important to notice that the county, Region Skåne, in its culture plan 2013–15 [Kulturplanen för 2013–15] a strategic development document concerning the region, specifically points at the "more opened" libraries as an important tool to further develop libraries as local meeting places. Although, Oldenburg himself, for good reasons, in 1989, had not thought of staff-less libraries as democracy enhancing meeting places, it seems to be recognized by cultural planners that open libraries can have a positive influence on the democratic development in local communities. The section below on "Social Capital" also deals with local democratic structures.

Changes of the Users' Behavior—Erving Goffman

Another interesting question addressed by the report concerns the behavior of the users. Does the absence of professional staff affect how the users use the library, for example, how they, without professional assistance browse and select among the different printed or other materials of the library? Library professionals who have criticized the staff-less library idea, often assert that users missing professional assistance are likely to make suboptimal choices, which could damage either their proper intellectual development or their ability to deal with practical problems. Therefore, the questionnaire contained a question whether the library usage during the staff-less hours differs from the usage during the staffed. The answers of the respondents confirmed that the user behavior was influenced by the presence of the staff. A total of 39.1% (9 out of 23) maintain that the behavior of the users is different during the staff-less opening hours compared to the staffed; 13% say "no" and 47.8% "do not know." One respondent remarks that users are more likely to choose controversial, for example, naughty literature when there is no staff present. Another respondent refers to a "greater freedom" when users are allowed to browse and select materials in private.

Such responses indicate that the presence of the staff contributes to maintaining certain social norms and behavioral standards. The Canadian–American influential everyday life sociologist, Erving Goffman (1922–82), who among others coined the concepts "backstage" and "frontstage," claimed that the behavior of individuals depends upon the presence of a "public" observer, which insists on the individual playing a specific role in certain contexts compared to a situation, where the individual is alone (Goffman, 1990; Johansson et al., 2015, p. 10). Such microsociological social mechanism and dynamics, revealed among others by Goffman, are also likely to be valid in staff-less library contexts. One question, however, is whether such social mechanisms are also present in staff-less library contexts too. Here, probably, most professionals would agree that they are present. Another question, which seems to be a much more discussed and controversial issue, is whether such a "greater freedom" and its associated dynamics and social interactions are desirable at all in public library contexts.

Serendipity in the Staff-Less Library—Lennart Björneborn (2008)

Will the users without professional assistance be able to find what they need during the unstaffed opening hours? How should the staff-less library be designed and furnished to optimize the users' opportunities for searching and finding the library materials they need? The answering of these two essential questions, which also occupies librarians responsible for staff-less libraries, are addressed in the report too (Johansson et al., 2015, pp. 21–22). The considerations in the report, especially, draw upon the work of Lennart Björneborn.

The Danish library researcher, Lennart Björneborn, basically, distinguish between two search modes:

1. the goal oriented, and
2. the explorative mode

The goal-oriented mode focuses on finding materials for specific purposes, which could be job-related tasks, study purposes, or hobbies, whereas the explorative mode is more determined by spontaneity and curious browsing. The user can shift from the one mode to the other many times during the same library visit. The big challenge for the library designer is, therefore, to have both search modes in mind when planning to organize the physical environment of a staff-less library.

Björneborn's research, furthermore, argues that there are a limited number of elements that have the ability to create serendipity in the library. The term "serendipity" denotes the ability to make unexpected, but, nevertheless, useful findings in the library. Björneborn defines, first, serendipity as a situation, where the user finds something that she had not expected to find or searched after actively. Second, serendipity appears when you find something useful or valuable in the library that you were not searching for, initially, but which later shows up to, for example, expand your horizon or in other ways introduces new dimensions into your life. People who have made strong serendipity-oriented findings, often wonder how they had been able to live without that particular dimension before.

Many principles and tricks can be used to increase the chance of making serendipity experiences in a public library. Open and inviting shelves seem to be among the fundamental requirements together with adequate signage, inspiring presentations, changing exhibitions, images, surprises, and the like. A respondent declares

> *I think that we've been better to make signs and clarify where things are located physically in the rooms of the library … we work hard to make the library available for the users even when we are not there*
>
> *Respondent—translated from Swedish by CGJ.*

How the users navigate around the library, therefore, becomes an important issue for staff-less library planning and design. Also the concept "information behavior" is something that should be taken seriously according to Björneborn (2008). To achieve research-based evidence, data have been collected through observations and interviews by Björneborn. Although Björneborn's study was not conducted at a "more opened" library site, it is likely that Björneborn's results are also relevant in staff-less contexts because the physical aspects and the furnishing are not significantly different through the staffed and unstaffed hours.

An interesting result shows that the information behavior of the users might shift many times from visit to visit dependent upon specific aspects of the life of the user. This individual information behavior variability makes it difficult to create stable user segments.

One of the consequences of the adaptation of diverse complicated library structures to staff-less contexts influences the common systems of classification in Sweden, the Swedish Library Classification System. Here, several adjustments and changes have been made.

The report concludes that Björneborn's work has been considered to be a valuable tool in the process of transforming design and furnishing and developing serendipity in Swedish more-opened libraries.

Library Anxiety (Eklof, 2013)

Typically, reports on open libraries will touch upon the question of safety: what has been done to ensure that the open library visitor feels safe. Here, even small architectural details that in daylight are totally nonproblematic can create anxiety and less usage of the library during the unstaffed hours. Questionnaires and interviews also confirm that a panoptical overview from outside the library is essential for the user's sense of security.

Technical Difficulties

The Swedish report (Johansson et al., 2015, p. 23), however, also deals with another, relevant aspect of library anxiety: the concern of many users for being unable to handle the technical procedures associated with entering and using the library. The staff-less library is, essentially, a high-technological solution where a certain amount of technical skills should be at hand by the users when entering the library and when using its facilities. This second aspect is all about how a technological unfamiliar user experiences and understands the grounds of the staff-less library space. It could, for example, be the practical aspect associated with computer-based literature search (Eklof, 2013, p. 250). Here, differences between age groups and generations can play a significant role. Inexperienced library users among young people are, maybe, more comfortable when using an interactive screen than members of elderly generations who prefer human assistance. Here, research shows that library anxiety should be handled quite differently because the interaction with the staff could probably also aggravate a young man's library anxiety (Eklof, 2013, p. 251). It is also a question asked by the report whether the need for staff-less library users to identify themselves to get access to the library could maybe induce library anxiety and prevent the use of the library during the staff-less opening hours for certain users.

It is also worth mentioning that the application of the staff-less model in Sweden, apparently also has influenced the feeling concerning safety of the staff. One respondent remarks:

in small libraries many librarians, for safety reasons, prefer to work in pairs in the evenings and when it's dark in winter, but I know that this is no longer the practice in Veberöd, because it is now felt that if the users can appear alone and feel safe, the staff should be able to do the same

Interviewee number 1—translated from Swedish by CGJ.

The Swedish report also considers other elements of importance for the user's feelings of safety, for example, the size of the library. Here, it is believed that larger spaces are more anxiety causing than smaller. However, comments from several respondents suggest that the widespread fear of being alone in an empty library is about to disappear (Johansson et al., 2015, p. 23).

The considerations made and the theoretical framework on "library anxiety" of the Swedish report are both very relevant and useful to enhance our understanding of the staff-less library model. We will now take a look at some of the broader, societal perspectives, social capital creation, and local community development dealt with by Johansson, Lindberg, and Eckerdal.

Social Capital (Vårheim, 2014)

The concept social capital has already been mentioned in Chapter 1. Social capital was also among the aspects that the authors of the Swedish staff-less library report have found interesting and relevant. In particular, there has been a focus on two concepts, "generalized" and "particularized" trust. "Particularized trust" concerns the propensity of people to form groups with people with whom they share some basic qualities such as age, family, sex, political beliefs, etc. "Generalized trust," on the other hand, concerns the ability to socialize with and to trust strangers with a background different from one's own. Especially, the "generalized trust," is believed by many researchers, e.g., the Norwegian library scientist, Andreas Vårheim (2014), to stimulate the formation of democracy. Vårheim's study builds upon the results of a survey conducted at a public library in the United States. Vårheim claims that the public library can contribute to the creation of generalized trust through the arrangement of activities like computer and Internet courses, different language cafes, and the like. Through the survey, Vårheim observed how the participants went from interacting with other course participants to interacting with other, previously unknown, library users. Hereby, a new community based on a common goal to utilize the library and to search for knowledge was created.

In interview number 2, the respondent describes a changed behavior of a user, influenced by the staff-less library, in the following way:

I feel that the users help one another much more than before ... there are always somebody who reacts and takes action when there are somebody who needs help ... in the past, I do not remember something similar

Translated from Swedish by CGJ.

Cases and examples where users of the staff-less libraries seem to interact more with each other than earlier appear to be available. The distinction between "particularized" and "generalized" trust is, however, important to be aware of in the future, too. Especially, it is a challenge to watch if the use of the staff-less library as a meeting place is primarily utilized by established group formations and joined companies or by newly established sociological and demographic constellations, which are more likely to produce "generalized" trust than established groups and clubs. The authors of the Swedish report here stress that it is not necessarily a task of the staff to set the agenda for meetings and other activities but rather to stimulate and facilitate the democratic processes, for example, through marketing and other efforts to facilitate the staff-less library as a dynamic meeting place.

At the same time, it is a rather controversial question among many library professionals whether the local democracy will thrive within such a decentralized and bottom-up managed framework. The question of releasing control is also important for the Swedish respondents. One of them mentions that the users in the staff-less contexts have become more independent and able after the introduction of staff-less opening hours. One respondent mentions that a user who earlier was able to solve a problem in the library now actually also does it. The increased do-it-yourself-capability, should, in principle, enable the staff to engage in new communication and dialogue activities toward their users.

Local Community Development—Casper Hvenegaard Rasmussen, Henrik Jochumsen, and Dorte Skot-Hansen

The important question whether more-opened libraries were likely to influence and how they did it, some rather abstract—but nevertheless important—relationships like identity, image, and the like in the local communities, was also addressed by Johansson, Lindberg, and Eckerdal. That public libraries, in general, can contribute to create a positive image of a town, a city, and an area has for long been recognized by several library researchers (Hvenegaard Rasmussen et al., 2011, p. 74; Jochumsen & Hvenegaard Rasmussen, 2000).

In one of the interviews, it was also emphasized that the "more opened" library was likely to create a positive image for a municipality and a local community. The respondent, here, referred to about 60–70 study visits from other municipalities all around Sweden as a sign of generation of a positive image of the county library in Lund.

The Library as the Only Cultural Institution

The role of the library as a cultural institution in the local community is also dependent upon the presence of other similar institutions. If the library is the only available public institution in the local community, its importance can be quite central. According to the Swedish survey, 60.9% (14 out 23) of the respondents indicate that the library is the only cultural institution in their local community. Hvenegaard Rasmussen et al. (2011) emphasize that the public library today is much more dependent upon its relationships and its partners than earlier. Indeed, such growing expectations are associated with a new role of the public library as an active participant in the emerging experience society. It seems as if the more-opened library in small local communities is particularly adapted to act as a kind of mediator, facilitating cultural, experience oriented, and meeting place activities. Due to its enlarged opening hours, the staff-less library is probably better to fullfill such a facilitator role than a traditional library with only few local opening hours. The point is emphasized in interview number 3:

> *We must change our activities radically to survive… and here availability is one important goal and an attractive content another*
>
> *Translated by CGJ.*

Cooperation and partnerships with local associations and organizations, as recommended by Skot-Hansen, Hvenegaard Rasmussen, and Jochumsen, have been implemented in many of the Swedish staff-less libraries, for example, in Veberöd and Glumslöv. Here, the library has strengthened its relationships with the local community by acting as a meeting place, a facilitator, and a bridge builder. In the cultural plan for the Skåne region 2013–15 the libraries are encouraged to improve and renew its working methods to become

> *more attentive in the local community outside the institution*
>
> *Region Skåne u.å, s.15—translated by CGJ.*

The respondents number 2 and 4 explicitly mention that both Veberöd och Glumslöv have groups who use the staff-less libraries as meeting places:

> *The Knitting Café was an excellent idea because when we started, we used to meet at downtown café's (…), but since then there have been many spontaneous meetings, if, for example, a group containing many people wanted to meet and talk; here, the library was an obvious place to come and sit down*
>
> *Interview number 2—translated by CGJ.*

In Glumslöv many local groups use the staff-less library:

> *the village community, and the antenna association have their meetings there and, in the autumn, some girls are going to start a reading circle there*
>
> *Interview number 4—translated by CGJ.*

Summary of Staff-Less Libraries—A Concept for All? (2015)

As to the first research question about the librarians work in a "more-opened" library, some traits are identified. The empirical data from the involved library leaders and staff indicate that the attitudes toward the more-opened library have been positive. The understanding of staff and library leaders about why the more-opened library model has been implemented also appears to be adequate. This suggests that the communication processes concerning the preparation of the "more-opened" activities have worked well. Although no strategical documents concerning the "more-opened" activities have been prepared, this deficiency has not hampered the staff in becoming familiar with the overall purposes of introducing the "more-opened" model in the Skåne libraries (Johansson et al., 2015, pp. 24–25).

As to the second research question about demographic and geographical characteristics of more-opened libraries in Skåne, the report concludes that a common trait, consisting of "rural areas with middle class families" could be identified. Furthermore, it was concluded that social control and a strong local anchoring ("bykänsla" in Swedish) were important for staff-less libraries to thrive. On the contrary, the confidence that staff-less libraries will work in more urban environments is less pronounced (Johansson et al., 2015, pp. 24–25).

More-Opened Public Libraries in the Experience Society

Concerning the use of the more opened libraries, the report assumes that the library sector is in a situation characterized by changing norms and shifting paradigms. This assumption is based on the observation by the respondents that the users seem to behave differently within the staff-less than within the staffed opening hours. This observation, furthermore, indicates that the presence of the staff contributes to regulate and maintain certain behavioral norms in the library. It is, furthermore, pointed out that the staff-less library potentially can act as shaper of identity where it is found. Especially, in areas where the public library is the only cultural institution it can achieve an even more central role in the local community. This central role is, furthermore, associated with expectations of the library as a central player in the emerging experience society. Such a role, however, requires that the public library adapts its activities in the direction of the upcoming experience society.

Finally, about the strategic aspects, the orientation toward an experience society library has already been mentioned. Besides, the local meeting place aims had been a prominent strategic purpose behind the staff-less library model. The idea of the library, including the staff-less library as a social capital enhancing unit in the local community, has been prosperous, especially the social capital produced, includes the sought after "generalized trust."

The 2015 report of Johansson et al., thus, deals with a number of relevant and interesting theoretical frameworks combined with a sound empirical basis. However, some burning questions are, for natural reasons, not answered. For example, whether a cultural institution that works well in agricultural surroundings is necessarily impossible in more urban contexts? We have seen that staff-less libraries have worked in densely populated urban districts in the Far East (e.g., Singapore, Taiwan, and more).

The next Swedish report to consider is the 2011 report on the first more-opened library in Veberöd, titled *"Thanks the more opened library there exists a library even for me."*

"Thanks the More Opened Library There Exists a Library Even for Me" (2011)

The Kultur Skåne (2011) report, in many respects, complements the 2015 Johansson et al. report. Especially, the voice of the users is more prominent in the 2011 report being based on both a quantitative web questionnaire and qualitative in-depth interviews. The data were collected about a year after the Veberöd open library started and the report is authored by people attached to public libraries in Skåne.

Veberöd is a small town or village with 5577 inhabitants in 2011. Web-questionnaires were sent both to users and nonusers of the library. Because the response rate was very low among the nonusers, we will here only focus on the user respondents. Here, the response rate was about 50% (questionnaires were sent to 575 and returned by 285).

About 90% of the respondents indicated that they had used the open library; 66% of them were women. Their activities during the unstaffed hours included loans and deliveries (90%), pick up of reserved materials (43%), and reading (26%). An important result was also that 93% of the open library users indicated that they felt safe during their library visits. Here, it should be mentioned that the feeling of safety was not safeguarded through the use of video surveillance equipment and CCTV. Indeed, the use of such equipment is very restricted according to Swedish law. Nearly all, 96%, of the respondents indicated that they wished to use the open library in the future too. A typical comment sounds

> *Fantastic! I think that the library has many good offerings … If the library had not been more opened, it had been difficult for me, because of my job, to utilize it before it closed. Since the library was made more opened, I've not visited it a single time during the staffed opening hour but nevertheless been able to use it twice a month.*
>
> *Kultur Skåne (2011, p. 5)—translated by CGJ.*

None of the comments, however, indicate that the users were dissatisfied with the staff or that the library also offered staffed opening hours. On the contrary, the report emphasizes, like the comment below, that it has been an important condition for the users that the employment of the staff was not threatened:

> *I was afraid that the more opened library was misused to reduce the staffed opening hours. It would have been a disaster! The two systems complement each other. On weekdays I return late and I can now, comfortably, borrow new audiobooks.*
>
> *Kultur Skåne (2011, p. 6)—translated by CGJ.*

The report also contains statistical data on visits and loans and also an interesting financial statement on both one-time costs and annual expenditures. It is remarkably although fully understandable that costs concerning monitoring technology are not among the items of expenditure. Totally, the expenditures in 2010 amounted to about half a million Swedish crowns (about $40,000) and $6000–7000 per year. The technical equipment and the software systems were installed by Axiell Scandinavia in cooperation with Bibliotheca. The log in procedure included two options: (1) a fingerprint-reader and (2) a traditional card reader combined with a PIN code.

Librarians Initiated Sweden's First More-Opened Library

Probably, the most interesting part of the report is, however, the qualitative part based on in-depth interviews and group interviews with staff members, leaders, and users. The author of the qualitative evaluation is Barbro Ross, the library director in Kävlinge municipality. She provides some interesting details about Sweden's first more-opened library. About its prehistory, she mentions the planning of an advanced research library in Brunnshög in the university town, Lund. Two public librarians, Karin Sandberg, library director, and Ingrid André, library information technology (IT) expert, then got the idea of developing an advanced public library solution. After a study tour to Denmark in 2008, they initiated and started up an open library project that was then accepted by the authorities in the autumn 2008.

Among the 11 districts in the municipality of Lund, Veberöd, in the Eastern part, showed to be the most interested. Its public library was situated in the central part of the town, surrounded by church, school, and shopping center in a new, attractive building. The necessary systems and equipment were installed by Axiell and Bibliotheca and the library opened in November 1, 2009. The interesting aspect is that the first open library project was initiated by librarians—just like in Denmark.

Citizens Concerned About the Staff

Nevertheless, a heated debate took place afterward demonstrating that the issue was considered to be highly controversial. Especially, the debate focused on the consequences for the staff. Would they become superfluous? It should also be mentioned that also the citizens of Veberöd were very concerned about the future of the staff; the critiques were not only found within the library profession. Such anxieties, however, among the users have apparently disappeared according to the results mentioned earlier of the quantitative survey.

The qualitative survey focuses on the following six problems:

- To what extent has the more opened library succeeded in meeting new needs and habits for readers and cultural consumption?
- To what extent has the public library of Veberöd increased its function as a cultural center for the local community?
- To what extent has cooperation between the library and local institutions, associations, and the like increased?
- To what extent has the image of the library among the Veberöd inhabitants been changed?
- What characterizes the use of the library during the staff-less opening hours compared to the use during the staffed?
- What characterizes the role of the librarian during the staff-less opening hours compared to the use during the staffed?

Kultur Skåne (2011, p. 13)

The applied methods were, primarily, in depth and ordinary interviews with the library director and staff (both librarians and clerical staff members), citizens (both users and nonusers), and local and regional politicians, system deliverers, and the like. The research design appears to be thorough and solid. Discourse analyses of the debates are, for example, used to identify relevant themes and debate issues for the interviews. Furthermore, a public debate at the library with 40 participating local users is part of the empirical evidence collected.

The evaluation also attracts interest because it documents an evolutionary trend often experienced in connection with open library implementation projects. In Denmark—and apparently also in Sweden—a remarkable shift of mood is often observed when the open library is introduced for the first time in a local community and when it is later implemented and people learn what its really all about. In the report the process is described:

> *In the beginning there was widespread resistance towards the project both among the staff and among the users. Why Veberöd? The library in Veberöd had already many opening hours compared to other branch libraries. Only a small amount of information was available for the staff. In was first after all the staff had been on a study visit to Denmark that the project became Veberöd's own project. Then the staff decided that the project should be successful. Our project and thinking big and innovative. It was also at this time that the staff realized that the library was not their library but belonged to the citizens of Veberöd ...Indeed, all the discussions about staff reduction, vandalism and theft influenced the users more than the staff and it took a long time to convince the inhabitants that it was a development that the staff believed in and that it was a development that the users would benefit from through a much more available library.*
>
> *Kultur Skåne (2011, p. 15).*

Three of the six key questions of the qualitative part of the survey, concerned new user needs, effects on the library's image and different ways to use the library. As to the needs it is not particularly surprising the needs of commuters and people with odd working hours represent an important background for the more opened library in Sweden. But it also seems that the prolonged opening hours also have changed the way the library is used. A user answered the question: "Do you use the library in a different way after the more opened offer was established?" in the following way:

> *I think the more opened library is a brilliant idea. I work away from the town and often I also participate in activities just after work. therefore the more opened opportunity means that I can visit the library later in the evening. I now spend more time at the library after it became more opened than before. Earlier, I often arrived precisely before closing time when the library was opened in evenings. Now I walk around for a looong (sic) time browsing the shelves. I often sit down and turn over the pages in books and journals which I never had time for earlier. The more opened library marks a boost for my loan of books and it is also more often than earlier.*
>
> *Kultur Skåne (2011, p. ?).*

The earlier quote delivers a good example of how prolonged opening hours can lead to both more and different library usage. The effects of the introduction of the open library were, however, also far more radical. It also influenced attitudes concerning the library among both the users and not at least among the staff. First, the image of the more-opened library as a permanently staff-less solution evaporated among both staff and users. Second, after the staff had started to act proactively and taken ownership of the project, they discovered that the project was not their project but belonged to the users of the library. Indeed, many staff members began to reconsider what their original attitude to their users really had been. Why had the fact that the users neither did destroy nor steal from the library been the most surprising event? This experience was to many staff members a true, attitudinal wake-up call.

The changing attitudinal positions were also found among the users. Indeed, a comparison of the results of the first quantitative survey and the qualitative interviews and the responses given to the associated questionnaire could document a remarkable attitudinal change. The change is interpreted by Barbro Ross as a consequence of the users having realized that the open library will not lead to staff reductions, but only to service improvements and increased availability.

Changed Staffed Opening Hours

Another consequence of the more-opened experience could be changed staffed opening hours. Some respondents have indicated that the library should not continue with its staffed opening hours unchanged. The users who are able to visit the library through the present opening hours, probably, do not make up a majority. Therefore, it would be better for the libraries to make their staffed opening hours fit to the habits and preferences of their users. Other public institutions already do. Museums, for example, keep open during weekends. The issue also led to the discussion about who should decide at what times the libraries should be staffed? The users or the library?

The report also compares the resistance toward the open libraries with earlier times reluctance toward new technology. An example could be the resistance against automated delivery and loan systems, and even, earlier, resistance toward the transition from handwritten to typewritten catalog cards.

One interesting conclusion made about open libraries is that their implementation has made evident that the users are ready to assume tasks that librarians really do not like. Barbro Ross is convinced that libraries can utilize the staffed opening hours more effectively both in terms of what they do and when they do it. This task, however, demands leadership, clarity, and with a staff group that feels confident about its jobs in the future.

Summary of "Thanks the More Opened Library There Exists a Library Even for Me" (2011)

Although many of the considerations contained in the Kultur Skåne (2011) report probably are less current now than they were 5 years ago, the report is still interesting in a number of respects. One important dimension deals with attitudes and how they can change. We do not know whether the same mechanisms work in all countries but the Veberöd case shows that the users can be very concerned about the well-being of "their" library staff. Users do not like open libraries being used to cut budgets and fire staff. On the other hand, the Veberöd case, however, does not document that users always prefer more staffed opening hours to prolonged unstaffed hours.

Perceptions and Realities are Two Different Things

Another principal lesson learnt is that perceptions about users and realities can be two very different things. The professional librarians apparently had low—too low—expectations as to their users' honesty and sense of responsibility toward the library and its premises. Trust in the users did not seem to have been the most prominent staff attitude in Veberöd.

A third point of the Veberöd experience concerns the more-open-library model itself. The case demonstrated that a staff-less library without CCTV equipment and video surveillance works. It is not an absolute necessary or a "sine qua non" condition.

The penultimate text is authored by Lisa Engström before she started her Ph.D. project in Copenhagen.

Just Staff-Less? (2015)

The article is based on a master's thesis containing interviews with users of four more-opened public libraries in Sweden. Quite recently, a more elaborated article has been published (Engström & Eckerdal, 2016). The differences between open libraries in Sweden and Denmark are emphasized in the article. Where open libraries in Denmark, and also, for example, in the United Kingdom, are associated with savings, the context in Sweden has been quite different. An interesting distinction between three different forms of availability are made:

1. time,
2. geographical, and
3. economic.

Especially, the first two forms play a role in more-open library contexts. The article concludes that the users appreciate the more-open libraries and also that the libraries can play a constructive role as a local meeting place. Finally, we will look at a report from 2011, which is also relevant, although it does not mention the more-open library concept with a word.

The report was prepared by the Swedish Library Association (Wagman, 2011).

Olika syn på saken Folkbiblioteket bland användare, icke användare och personal (2011)

Although this report does not specifically mention the open-library model, the topics dealt with are much related to the issues treated earlier. The report contains the results of a user survey based on a random sample of 800 users and 100 staff members. Nonusers are defined as respondents who have not visited a public library the last 4 years and frequent users have visited the library at least every half year. Nonfrequent users lie in between. The reported is written by the Swedish Library Association's researcher, Anna Kåring Wagman.

The frequent users were asked which activities they performed when they last visited the library within the last 6 months. Here, between 30% and 35% of the female users put a cross with the option "Asked the librarian for help," whereas only 25% of the male users did the same. This result is also interesting in an open-library perspective because similar surveys in Denmark reveal small reductions as to how much and for what purposes the staff is used.

The Most Important Services are All Staff-Related

The Swedish respondents were also asked what they, among 30 different options, considered to be very important services in a public library. Here, the top three preferences were all staff related: a competent and knowledgeable that makes an effort and makes you feel welcome. First, on the fourth place, we find an option that points at the more-opened model: it is important to me that the library has convenient opening times. This option was chosen by about 60% (Wagman, 2011, p. 11). Apparently, the Swedish public library users seem very much to appreciate the services of the staff. We find a similar appreciation and prioritization of the staff in the Veberöd report.

Lesson Learnt—Be Careful About User Surveys!

However, it can be dangerous to base and deduct library service development only on user surveys. According to a thought experiment, political decision makers and library professional could claim—based on Wagman's (2011) survey, that the open-library model impossible would fit into a Swedish public library context with its documented tight relationships between users and staff. A catastrophic failure caused by ignoring the available existing evidence! Maybe the model would work in Denmark but Sweden is something quite different. Nevertheless, things turned out somewhat unexpected. The faithful users in Veberöd at the same time both demonstrated their recognition of the staff and began to love and use the staff-less library. The lesson learned from this incidence is that innovations never tested, sometimes ought to have a chance to be tried out instead of being rejected based on apparently obvious reasons and deductions.

Gaps

Wagman (2011) gives rise to another open-library relevant point. After the analysis of the preferences and priorities of the users, they are asked to indicate to what extent they agree with a number of claims about the library. Then the staff respondents are presented to the same claims and the answers were then compared (Wagman, 2011, pp. 16–20). The analysis reveals a number of interesting discrepancies between the perceptions of the users and the staff. An illustrative example shows that 92% of the staff totally agreed with the claim that their users feel welcome in the library, whereas only 58% of the users have the same experience (Wagman, 2011, p. 20). However, concerning the agreement to the claim about adequate opening hours, users and staff seem to agree more or less: 40% of the staff fully agree that their opening hours are appropriate, whereas 35% of the staff do fully agree (Wagman, 2011, pp. 17–19).

The analyses of the different gaps between what the staff thinks are important for the users and what the users themselves feel are certainly also highly relevant in open-library contexts whether they relate to attitudes on honesty and moral habitus or to more innocent issues. As we will see, the negative staff expectations concerning bad user behavior is often much worse in some Central European countries than in the trust-oriented countries within the Nordic hemisphere.

Other Gaps—Now Between Bureaucrats and Users

Another dimension of attitudinal differences than the user–staff distinction concerns the gap between local politicians and their officials and bureaucrats and the library users. Here, Wagman (2011, p. 22) refers to earlier surveys by the Swedish Library Association, which compared the distinctly modern perceptions of the library as an actor within an upcoming experience society, shared by 85% of their segment, and the more conservative perception of the users where only 45% share the high flying visions of the experience society. Here, in the present contexts, it should be added that the politicians and bureaucrats also were highly in favor of the-library-as-a-meeting-place vision. On the other hand, whereas 85% of the users preferred a traditional library with books and a calm and silent atmosphere; only 15% of the local politicians and bureaucrats were in favor of that vision.

We will now leave the Nordic countries and turn to the United Kingdom as the last country. Indeed, there seems to be no comprehensive reports on open libraries in Germany and the United States.

The United Kingdom

Review of the Open Library Concept in UK Public Libraries—Report for Bibliotheca (2016)

The 34 pages report is written by Maria Kelly on behalf of nonprofit firm, APSE (Association for Public Service Excellence) with its main office situated in Manchester. The report presents a review of eight of the nine British councils who have implemented open libraries. The review took place throughout May–July 2016 so the data in the report are quite current. Typically, the eight councils have only implemented or piloted the open libraries in one of the public libraries within their districts. The reviewed eight councils are London Borough of Barnet, Devon County Council, London Borough of Harrow, Hertfordshire County Council, Norfolk County Council, Peterborough City Council, Stockport MB Council, and Trafford MB Council. The reviewed open libraries are often found in busy urban environments.

A Useful Framework

The report is, especially, distinguished by its distinctly systematic approach. Each of the open library projects are described according to the same structure which makes analyses and comparisons easier. The structure consists of the following framework and elements:

- Cost components
- Benefit components
- Saving areas
- Lessons learned
- Challenges overcome and areas of good practice
- Future plans

"Cost components" include the capital costs and the costs for ongoing support. Benefits could, for example, be integration with the LMS, the extended opening hours, increased number of users, new user groups, and changed staff attitudes. If the staff, for example, act as ambassadors for the open library project it is, naturally, considered to be a benefit gained. In the review of open libraries in London Borough of Harrow we, for example, learn that "Extended hours using the Open Library Concept at Wealdstone (rose) from 40.5 staffed hours to 40.5 staffed hours plus 17.25 unstaffed hours" under the headline "Benefit components." Under "Savings areas" are the information brought that "The Open Library Concept was "part of a package of savings of Xt £'s" and that staff savings were achieved from the "closure of… four" libraries. There were no staff savings from the implementation of the "Open Library Concept at Wealdstone." "Lessons learned" could be many different things such as that teething problems happen and that it takes time to get staff on board. "Challenges overcome and areas of good practice" could, for example, be

- "A project manager would have been useful"
- "Buy-in from Elected Members—they need to see it working"
- "One size will not fit all—each building is different"

Finally, "Future plans" which deals with—yes—future plans.

A framework like the one presented above has many advantages. Besides facilitating sharing of information and making comparisons easier, it can probably also help to clarify positions and debunk myths.

Values Associated With Open Libraries

The reports deals with a number of different values that are activated or strengthened along with open library implementation. Increased usage is already known from many countries but also the "potential impact on social inclusion" is recommended many times in the report as a value particularly furthered by open libraries along with phrases like "libraries are trusted spaces." It is often mentioned explicitly that the unstaffed opening hours are used for meetings and social events. Social inclusion is a value akin to concepts like "social capital" and "low-intensive" meeting places. The idea of the open libraries as a means to "give the library back to the community" is also prominent in the report.

Recommendations

The report presents a number of—in total 16—recommendations. It is, for example, emphasized that during the decision-making phase it is important if not essential that the key stakeholder see themselves the system in action. The report also recommends what they call a "phased introduction" with a pilot project followed by further reflections on the lessons learned. To understand the user needs, recommendation number 7 suggests a series of consultations even on every stage of the project with the public and with community groups to understand their needs. Indeed, it seems as if this recommendation has been followed by many of the involved councils. A consultation with users in Peterborough thus revealed that among the three options below, it was certainly option number 3 that was supported by the users:

1. Close all local branch libraries and invest in the Central Library.
2. Keep all libraries open and reduce opening hours across the service.
3. Consider new ways to deliver the services to maintain and extend opening hours.

Several consultations, furthermore, have revealed that, for example, 75% of the respondents thought that access to outside of the normal opening hours was important.

Also engagement of the staff is considered to be essential. The staff has a key role, for example, by promoting the open library, by inducting new users, and by designing new procedures. Like one of the Swedish more open library reports, Kelly must also admit that

> *where assurances can be given or there is no impact on staff hours this clearly helps engagement.*

Also like Sweden "the public values local staff" are emphasized. In four of the councils, nevertheless, staff cuts have taken place. The report also mentions that in some councils staff has launched industrial action against the loss of job and as they see it, a reduction in service. It is explicitly emphasized by Kelly that the review will also cover this issue. Staff are also characterized as apprehensive and worried personally for both their job security and for what might happen to the library (vandalism, misuse, safety) when it becomes an open library. It is also underlined that staff in other countries, for example, in Denmark had similar concerns regarding vandalism and misuse. It is also recommended that an individual from the library service should be appointed as project manager and that sufficient time (3–6 months) should be spent in the planning phase.

Indeed, economy and financial consideration seem to play a significant role, and it is argued that the

> *driver for new ways of working [in casu: the open libraries. CGJ] is often the need to deliver services within a reduced budget.*

Finally, it should be mentioned that the report recognizes that open libraries are both a "complementary and integral part of an overall library service" and that the successful model has been pioneered and extensively used across Scandinavia since 2009. Thus there is no hint in the report that staff-less libraries are likely to make the traditional staffed library superfluous.

We will now take a look at the key open library stakeholders.

The Role of Stakeholders

The English Kelly (2016) report, referred to in the previous section, operated with four key stakeholders:

- elected members,
- staff,
- community, and
- users.

An extended list of nine relevant, principal stakeholders is presented below:

- Users
- Local politicians
- Newspapers, net media, and the like
- Government agencies
- Interest organizations
- Trade unions
- Professional and clerical staff of the library
- Library leaders
- Library researchers

In this section we will look at the roles, attitudes, and activities of the different stake holders,some of their characteristics, and some interesting national variations. We will start, as a point of reference, with the roles played by the Danish staff-less library stakeholders since 2004.

Users

Here, the voice of the user was listened to when the first open libraries emerged in Denmark and Sweden. In the United Kingdom, the so-called user consultations are, typically, conducted even before pilot studies were launched. Later, user surveys have also counted and calculated the user satisfaction with the open libraries. The typical results of both the systematic user surveys and anecdotal stories were excited users. It should be mentioned here that the great user satisfaction rates with the open libraries are not at all associated with dissatisfaction with the staffed services offered in the library.

Local Politicians, Elected Member, and Media

The role of the local politicians is often more powerful than the users. However, their attitudes pull in the same direction. Often, local politicians prefer to protect and develop small local communities and use staff-less library as a tool to obtain such purposes. Local newspapers, usually, cover the opening of new open libraries with excited reportages. Only in certain national newspapers have critical positions been voiced. Few web sites have been devoted entirely to cover staff-less libraries (Box 3.4).

Government Agencies

Government agencies in Denmark like Kulturstyrelsen, the highest library authority, has actively supported the open libraries through project support, positive attention, and different support statements in different media. The former director of the Danish Agency for Libraries and Media, Jens Thorhauge, has on several occasions supported the open libraries.

Box 3.4 Cross-Party Support in Peterborough, UK—Case

Results: Lisa [Roberts. CGJ] was delighted to summarize the solution as, "giving library users everything they asked for," and explained, "open+ has been really well received by our cabinet members, with cross-party support. The solution has allowed us to make the required saving, keep all of our libraries open, future-proof our service for the next 5–10 years, and extend opening hours by almost 50% across all sites.

Open+ brochure, issued by Bibliotheca (2016).

Interest Organizations

Interest organizations within the library field such as the Danish Library Association have also supported the open libraries, although certain conditions have been articulated as decisive. Especially, the broader context of the implementation has played a role. If the establishment of an open library leads to no staffed opening hours at all, the library association declared that it would be against an open library model. The optimal case occurs where extended opening hours are established and the existing staffed opening hours kept intact. In Sweden, for example, the library association has facilitated the debate between open library pros and cons.

Staff

Compared to the library interest organizations, the attitudes of the professional trade unions were much more critical. Skeptical is certainly the adjective that characterizes the attitudes of the Danish librarian's trade union best, whereas the staff trade unions in the United Kingdom and Ireland sometimes have launched industrial action against open library projects.

For historical reasons, there seems to be differences although small between librarians and clerical staff as to staff-less libraries. When automated delivery and loan systems were introduced in the 1990s, the clerical staff were much more concerned than the librarians. The open libraries, on the other hand, now seem to affect the librarians more than the clerical staff. However, both the clerical staff and the librarians include supporters as well as opponents.

Leaders

The same can be said about the library leaders. Many of them are keen supporters of the staff-less libraries with excited reports in interviews, newspapers, and professional journals. Other leaders can be more lukewarm supporters, especially, when their staff-less branch libraries have been less successful in terms of visits or loans or for principal reasons.

Researchers

Researchers have only dealt with the open library issue in Denmark, Finland, Sweden, and the United Kingdom. Most of the contributions have been on the positive side emphasizing open library benefits such as its importance for revitalizing local communities and giving busy commuters access to public library services. Only a few critical voices have emphasized that the importance of the voice of the professional staff had been underestimated (Kann-Rasmussen & Tank, 2016).

However, also the critical points of otherwise open-library-positive researchers have been used by skeptical trade union writers. The Danish librarians' union, for example, has emphasized that Johannsen in his 2012 report observes that the number of visits and loans per hour during has always been much higher during the staffed opening hours than during the unstaffed hours (Lerche, 2012a).

In the ongoing discussions about open libraries among the different shareholder, the question of legitimacy plays an important role. I will, therefore, spend some words on clarifying the complex concept.

Legitimacy and Legality

First, it is important to distinguish between "legitimacy" and "legality." The legality of staff-less libraries has, usually, not been questioned apart from, for example, Finnish critiques who claim that staff-less libraries are contrary to the library law because the law as they see it requires permanent presence of staff (Mustikkamäki's, 2015, p. 21).

Legitimacy

The legitimacy of the open libraries is, however, a more complex question because it involves different interpretations of the concept. What it really means when you claim that staff-less libraries are legitimate—or the opposite—very much depends on how you define legitimacy. Therefore, it makes sense to offer the question at least some attention.

Suchman's Forms of legitimacy

The American professor, Mark C. Suchman (1995), has tried to synthesize the large literature on organizational legitimacy, highlighting similarities and disparities among the leading approaches. His analysis is clarifying by identifying three primary forms of legitimacy:

- pragmatic, based on audience self-interest
- moral, based on normative approval, and
- cognitive, based on comprehensibility and taken-for-grantedness

According to Suchman legitimacy is defined as

> *Legitimacy is a generalized perception or assumption that the actions of an entity are desirable, proper, or appropriate within some socially constructed system of norms, values, beliefs, and definitions.*
>
> *Suchman (1995, p. 574).*

Here, the "actions of an entity" could, for example, be the actions of a certain library or the actions of all public libraries in a specific county or country. The words "perception" and "assumption" refer to the subjective side of legitimacy. It is what users, professionals, audiences, stakeholders perceive or assume as legitimate that counts—not an objective standard of quality and usefulness. The term "socially constructed" refers to the circumstance that legitimacy is dependent on: (1) a collective audience or group of stakeholders and (2) accordance with the shared beliefs of that audience or group.

It means, for example, that open libraries could be considered legitimate by users, central authorities, and local politicians because the service corresponds to the norms and values of these groups, and—at the same time—less legitimate by certain library professionals because staff-less libraries are contrary to their definitions of what constitutes desirable and proper public library service. As a quote reproduced elsewhere a library without staff is not a library in the same way as an aspirin in a tin outside a house is not a hospital (see Chapter 1). Here open libraries are certainly not perceived as legitimate because they do not deliver appropriate public library services.

But the way different groups legitimize a service could also vary a lot. A user with odd working hours could probably rest staff-less library legitimacy on self-regarding, individual utility, whereas a proponent of local democracy and social capital would build legitimation upon larger cultural and political norms and values. I will now try to illustrate with examples from the open library literature some different applications of the Suchman's three forms of legitimacy.

Pragmatic Legitimacy

Pragmatic legitimacy is different from the two other forms, moral and cognitive, because it rests entirely on audience self-interest. It is, furthermore, interesting that pragmatic legitimacy is based upon cost–benefit appraisals through public discussions. It also means that organizations and stakeholders can win pragmatic legitimacy by participating in debates and dialogues. In staff-less library contexts we have had only few examples, among other things, from Ireland, where opponents among professionals have tried to involve the library users in their fight against open libraries. Mostly, however, the discussions about legitimacy have taken place within more or less closed professional circles and media.

Suchman identifies a related type of pragmatic legitimacy, which he calls "influence legitimacy" (1995, p. 578). Here, organizations show willingness to relinquish some authority to the affected audiences. This particular type is relevant in library contexts too, where many staff-less libraries try to establish partnerships with local associations and institutions concerning the use of the library's premises for meetings, work-shops, and other purposes during the staff-less opening hours. In the United Kingdom, for example, we find often public consultations and user surveys prior to open library project implementations.

Moral Legitimacy

Moral legitimacy reflects a "positive normative evaluation" of the organization and its activities (Suchman, 1995, p. 579). Unlike pragmatic legitimacy, moral legitimacy rests on judgments about whether an activity is "the right thing to do." Pragmatic legitimacy focuses more on whether a given activity benefits its users. Typically, moral legitimacy deals with whether an activity promotes social welfare or human rights. When you emphasize that staff-less libraries supports the creation or maintenance of social inclusion in a local community, you actually refer to a moral legitimacy argument. You may say that open libraries in rural environments are more likely to be legitimated through moral legitimacy than open libraries in urban environments. However, although the narrow interests of members of the so-called "creative class" are served by urban staff-less libraries, economic development and innovation oriented arguments provided by, for example, Richard Florida (2005) may also emphasize the societal values of the creative class. This emphasis, in turn, may contribute to transform the narrow self-interests of the creative class into proper societal values. Therefore, as Suchman (1995, p. 584) emphasizes, both pragmatic and moral legitimacy "co-exist in most real-world settings."

Cognitive Legitimacy

The most difficult of the three types of legitimacy dynamics, the third type, cognitive legitimacy, is based on "taken-for-granted-ness." Suchman considers this kind of taken-for-granted-ness as the most "subtle and the most powerful source of legitimacy identified to date." "If alternatives become unthinkable, challenges become impossible," the legitimated entity becomes nearly unassailable (1995, p. 583). This type of legitimacy generally lies beyond the reach of but the most fortunate institutions.

In library contexts, services like open shelves, children's departments, and fiction literature, probably, in many countries at least, have an almost taken-for-granted-ness status. It is, however, also obvious that open libraries have not yet achieved this kind of cognitive legitimacy.

Gaining, Maintaining and Repairing Legitimacy

Suchman furthermore distinguishes between strategies for gaining, maintaining, and repairing legitimacy of each type. Because open libraries are a relatively new invention, it is probably the gaining of legitimacy that is the most interesting activity. Suchman here presents (1995, p. 587–593) a number of strategies. He distinguishes between three different legitimacy-building strategies aimed at different audiences such as the existing users and new audiences. It is interesting here that library directors and national authorities often have focused on whether staff-less library offerings were likely to attract new categories of users. Another strategy found in library contexts is to strive toward serving the needs of already existing users. Here, library directors as well as national authorities have been eager to produce statistic figures on primary library performance indicators such as visits and loans to gain pragmatic and moral legitimacy for the open libraries.

Because the audiences of staff-less library legitimacy campaigns are often internal (the staffs or in particular resistant staff-groups within the library), the directors and the other proponents of staff-less libraries emphasize the building of a "winning coalition of believers" whose "conceptions of socially desirable activity set the terms for subsequent moral debate" (Suchman, 1995, p. 592). In a specific library context, it means that it is a crucial legitimacy gaining activity to convince all staff members that open libraries are socially desirable and in accordance with the basic goals and values of the public libraries in society. Many library directors highlight this task as particularly important. In the United Kingdom and in Sweden, the engagement of the staff is also considered an essential task in an open library project.

Summary

The aim of this chapter was to address the following four research questions:

- What characterizes the development of open library technological infrastructure?
- What characterizes the themes and findings of reports and research about open libraries in five different countries?
- What characterizes the different reactions of the principal public library stakeholders (e.g., local politicians, users, professional staff) upon the open library development?

As to the first question, the technological infrastructure and its key elements was presented. The specific importance of the applied RFID technology was explained and the importance of integrated solutions underlined. Finally, a number of innovative developments under the headline The Transformative Library demonstrated how the library could automatically adapt to the visiting users' wants and needs.

The findings of the research-oriented open library reports from five countries (Denmark, Finland, Norway, Sweden, and the United Kingdom) covered many different themes and topics. Many of the reports contributed to a description and also an understanding of both user attitudes and behavior toward the open library model. Although none of the studies documented that the users were dissatisfied with the services provided by the library staff members, the users were also very satisfied with the open library services and it is very difficult to find evidence that the users prefer more staffed opening hours toward many more staff-less opening hours. On the other hand, the users, especially in Sweden, dislike staff members being fired because of open libraries.

As to the principal stakeholders, it seems as if a "winning coalition" of users, local politician, the media, and library leaders in favor of open libraries faces a small group of open library resistant librarians. Furthermore, it seems that the segment of opponents is strongest in the pre- and planning phase and dwindles more or less when the open library systems become operational.

Open Library Communities, Users, and Usage

4

Introduction

The discussions concerning staff-less libraries very often include miscellaneous assumptions and hypotheses about what the user really needs as to library services. Do they really prefer prolonged opening hours to the permanent presence of professional assistance? In the preceding chapter we have already addressed that question. In this chapter focus will be on the users, the primary user segments, and their usage during the unstaffed opening hours.

Vandalism

Here, concerns about vandalism and similar risks were most prominent in urban districts. Indeed, there seemed to be no limits for what would happen, when the staff-less libraries entered the urban environments. In some cases, the concerns were well-founded, but mostly the gloomy prophecies remained, fortunately, unfulfilled. Today, open libraries are found in many Nordic cities; even in the capitals like Copenhagen and Oslo vandalism and theft have rarely or never occurred.

Undue Anxiety

Open libraries in urban districts had, however, also been successful far away from Europe. Maybe, experiences from densely populated, urban environments in Far East Asia (China, Japan, Singapore and Taiwan) could have reassured the concerned European professionals, that vandalism and theft are less widespread than often believed.

Causes of Vandalism

Besides urbanization, the chapter will also take a closer look upon what has led to problems with vandalism in certain branch libraries. Unlike Chapter 6, where we will look upon different strategies to counteract vandalism and the like; this chapter in the section on neighborhood characteristics will discuss to what extent it is possible to predict vandalism and thereby avoid installing open libraries in high-risk neighborhoods.

Open Libraries Designed for the Creative Class?

The section on sociodemographic factors will also consider which role different sociodemographic and life-style oriented factors play in open library contexts. Is it so that the open library model, in particular, mostly satisfies the interests of in advance privileged

Staff-Less Libraries. http://dx.doi.org/10.1016/B978-0-08-101923-8.00004-6

categories of users at the expense of weaker user segments? Are such assumptions evidence based or mostly based on loose hypotheses and social prejudices?

Urbanization

When the experiences with the more-opened model were evaluated in Sweden, one of the main conclusions was, that the more-opened model, especially, fits into

> *rural areas with middle class families*
>
> *Johansson, Lindberg, and Eckerdal (2015, pp. 24–25).*

We will start with the urban–rural distinction and return to the social middle class dimension in the section below on sociodemographic factors.

Differences Between Rural and Urban Areas

Here, we will focus on the distinction between rural and urban: why do open libraries apparently better fit into rural than into urban environments? According to common sense reasoning, rural settings are, usually, peaceful areas where everyone knows all and where vandalism and theft, therefore, rarely occur. More sophisticated explanations point to concepts like "social control" and a "strong local anchoring" to explain why more-opened libraries thrive in rural areas (Johansson et al., 2015, pp. 24–25). In Box 4.1, Finn Petersen reports on experiences with open libraries in the Danish capital, Copenhagen.

It is, however, somewhat misleading to imagine the local communities in Southern Sweden with more-opened libraries as traditional rural areas with settled peasants who spend their entire life in the countryside around their idyllic farms and villages. On the contrary, among the most eager users of the more-opened libraries we find busy commuters working in urban centers away from where they live.

Box 4.1 demonstrates that urban users too appreciate the open library model and that the challenges can be overcome. Concerning the rural areas, it is, however, a question which aspects of the areas that best explains their particular open library preferences. Indeed, traditional peasants in both Denmark and Sweden are relatively rare library visitors and the emergence of the open libraries has, certainly, not changed this particular social pattern. Both experiences from Sweden, the United Kingdom and the United States (Yacolt public library) suggest that it is the presence of modern commuters living in rural areas that explains the success of open libraries and not so much references to traditional, peasant lifestyles.

In the United Kingdom, for example, a case study shows how the open library functions in the small market town, Cullompton, in Devon, a town that largely acts as a dormitory for people working in the nearby cities. In Cullompton, many people are normally working elsewhere during regular daytime opening hours. The library director has observed how the core users have begun to take ownership during the unattended opening hours and also that the open library is used for meetings by 20–30 regular evening meeting participants (publiclibrariesnews.com).

Box 4.1 Open Libraries in the City of Copenhagen—Case

"Copenhagen Libraries opened its first unstaffed library in 2012. Then, two libraries opened with unattended access in the morning and again in the evening. During the day, however, the libraries were manned as usual. In 2016, this model was spread to all Copenhagen public libraries.

There was, of course, much focus on whether the unstaffed access was possible in a big city with all the problems unstaffed access to a public library might cause.

In general, the staff-less libraries have been also been successful in Copenhagen. Certainly, there have been challenges and problems; it has also been necessary to shut down the staff-less access in some of the more exposed libraries for a period. The most common problems have been groups of young people who "conquers space" inappropriately and also with the homeless. Most important, however, is that people are generally very happy with the opportunity to use their public library during a much longer period than previously.

On a scale from 1 to 10, user satisfaction has been measured to an average of 9.2. This knowledge we have from a user study, conducted in the spring 2016. The study was based on more than 1000 responses. Moreover, the same survey showed that the greatest use of the unmanned hours was loans and returns. 95% of the respondents reported loans and returns as their main purpose of the visit. The second largest user group indicated that they used the library's facilities for children. A third interesting result was that far more men than women stated that they use the unstaffed opening hours. 33 percent were men compared to only 20 percent women.

One last interesting trend, I want to mention is that far more people with origin different than Danish indicate that they use the library more during the staff-less opening hours: 34 percent compared to only 22 percent of Danish origin.

The most obvious challenge in the metropolitan area is certainly to ensure local ownership of the open library. This is one reason why the Copenhagen libraries in its library strategy "Empower Citizens, 2014–2019" focus on the need to stimulate organized user–driven activities in the library space when the library is self-serviced.

To sum up, one can say that although self-service access is not unproblematic, it works in general also in a metropolitan area and people love it!"

By Finn Petersen, head of Library Development, Copenhagen.

Neighborhood Characteristics

Today, staff-less libraries have been introduced in many different types of both rural and urban neighborhoods. In most of the cases the problems with vandalism and unrest have been minimal. In the few cases, where problems have occurred, it has often been difficult to point out, precisely, what aspects of the neighborhood that had been the underlying cause of the trouble. In the Danish capital, Copenhagen, for example, one urban branch library, Blågården's library, has been particularly plagued by problems such as vandalism and threats, whereas the neighborhoods of other open branch libraries, also in urban districts, have been quite peaceful. The same is true concerning another Danish town, Aalborg, with an urban population of 112,194 inhabitants in 2016. Here, some urban open library branches experienced serious problems, whereas other urban districts around have avoided them. However, not only open libraries in urban districts have had problems with unrest, but also rural district branches have experienced unrest and vandalism. In Chapter 5, we will look at what can be done and had been done to counteract unrest and vandalism.

Can Vandalism Be Foreseen?

But one could ask whether it would have been better much earlier to have foreseen the disturbances and avoided it so to say in advance by not placing open libraries in such risky districts? Immediately, a reasoning like that appears both to make sense and to be reliable. However, the history of the open libraries in, for example, Denmark and Sweden have often demonstrated that anxious staff concerns about vandalism and unrest have turned out to be grossly exaggerated. A timid and suspicious open library policy in Denmark and Southern Sweden would, certainly, have meant that a number of today healthy open library projects had never been implemented. Furthermore, open libraries might have other beneficial local community contributions similar to, for example, iconic architecture.

Iconic Architecture Can Lift a Neighborhood—Can an Open Library Do the Same?

It is a key element of some recent urban development strategies (Skot-Hansen, Hvenegaard Rasmussen, & Jochumsen, 2013) that spectacular buildings and avant-garde architecture can, so to speak, boost a worn and downgraded neighborhood. Indeed, three different culture-led urban regeneration strategies have been identified. The three strategies have been labeled as

1. cultural icons and landmarks;
2. place making and urban identity; and
3. community vitalization (Skot-Hansen et al., 2013, p. 9).

It is probable that the opening of a staff-less library may have a similar effect upon community vitalization as the opening of an iconic library building because the open library also sends a clear message to the outside world that central authorities have enough trust to develop this particular neighborhood. The open library can here be seen as an empowerment enhancing step. Open libraries are also recognized for facilitating improved social inclusion (Kelly, 2016). Showing confidence is likely to support the creation of local community social capital. It is also probable that the establishment of an open library in a district might have an effect that is akin to the effect of landmark, library architecture. The content of Box 4.2 emphasizes the importance of distance and opening hours for small local communities.

Box 4.2 Library on the Doorstep, Lyngby-Taarbæk Case (Denmark) Case

"For Taarbæk Library, the solution has meant that they can still offer residents in the community access to library facilities," Jeppe Bjørn, Head of Libraries in the Lyngby-Taarbæk municipality comments: "the budget cuts we were undergoing made it impossible for us to keep our local libraries running as they were—we needed to do something. Open+ has ensured that residents in the Taarbæk area still have a local library they can visit—they do not have to make a long journey into the City when they have this on their doorstep."

Since deploying the system, Taarbæk Library has been able to provide access to the library's facilities until 10 pm each evening, and from 7 am on the weekends – a vast increase on its previous operating hours of 2–5 pm on weekdays and 10 am to 2 pm on weekends. Taarbæk Library has been scheduled so that the library is unmanned the majority of time. However, the municipality has ensured that a certain number of hours a week are manned by staff, meaning if a patron does want to speak to a librarian they can arrange their visit to coincide with these hours.

Open+ brochure issued by Bibliotheca (2016).

Sociodemographic Factors

Especially concerning certain sociodemographic factors, there have been made a number of hypotheses, mostly among the professionals themselves, about which social groups' or segments' interests the open libraries especially serve and about whose interests' they, in particular, do not satisfy. Some antistaff-less library Internet sites or blogs even claim that the open libraries are designed and tailored, primarily, to serve the busy commuter and other members of the so-called creative class at the expense of people with disabilities and computer illiterate users.

Difficult to Find Out About the Social Background of the Open Library Users

Because the log in procedures only deliver identity (ID) data on gender and age and not on, for example, occupation and education, it is difficult to find out whether such claims are more or less valid and evidence based. Besides, user data are, usually, not kept for a long time. Consequently, we most often only have access to data on basic characteristics like age, sex, and the arrival time. It is, for example, possible to find out that the most frequent visitors of a certain open library during a certain month are people above 65 years and also that men make up 60% of the visitors. On the other hand, we do not know whether the plus 65 retirees also consist of most men or whether the percentage of men is evenly distributed on all age groups. The widespread hypothesis that younger males with small children are especially frequent visitors is thus difficult to either verify or falsify.

Valuable Data From Online Questionnaires

Luckily, some surveys drawing on other sources such as online questionnaires may complement the insufficient library use data. A representative Danish survey from 2014 (Moos-Bjerre & Moos-Bjerre) may, here, deliver some indirect hints. It should be mentioned here that the Moos-Bjerre (2014) survey does not operate with a classical sociodemographic classification scheme, but rather with a modern segmentation model based upon a combination of life-style features and sociodemographic traits (Johannsen, 2014b).

One prominent finding of Moos-Bjerre is the identification of the primary interests of the Danish library users. According to the user responses, the three primary library interests of the users are (Moos-Bjerre & Moos-Bjerre, 2014, pp. 80–97):

- Books (loan of) (77%)
- Flexible physical access (prolonged/self-serviced opening hours) (29%)
- Loan of music compact discs (CDs) and films on digital versatile disc (DVD) (26%)

These figures, however, do not tell anything about the different priorities and interests of specific social and demographic groups or segments. To simplify the comparison, we will here primarily concentrate on two of the five constructed plus 30 years segments:

- The cultural super user
- The mature lower middle class user

The "Cultural Super User" Segment

The "cultural super user" is, typically, represented by a middle aged (50–69 years) woman, highly educated and overrepresented in the larger cities. Her primary library interests are borrowing books (79%) and flexible psychical access (31%). A library service like computer and IT courses, on the other hand, only interests a few (three percent) of the members of this segment; 44% of the "cultural super users" have used the library within the last month (Moos-Bjerre & Moos-Bjerre, 2014, p. 50), so they belong to the most frequent library users.

The "Mature Lower Middle Class" Segment

The "mature lower middle class" segment is, typically, also represented by a woman between 50 and 69 years old. Both the income and the educational levels of this segment are, however, relatively low. Their primary public library interests are books (83%) and flexible physical access (24%). The "mature lower middle class" is a typical nonuser segment where only 21% have used the library within the last month. At the same time they represent the largest single segment in the population—18% (Moos-Bjerre & Moos-Bjerre, 2014, pp. 57–58).

The two segments just mentioned in many respects represent extremes among the Danish library users both in terms of library usage and in terms of educational background. Nevertheless, there are also apparent similarities between the two segments.

Admittedly, more members of the "cultural super user" segment are interested in flexible physical access than the "mature lower middle class" but the difference is not particularly high, respectively, 31% and 24%.

Appreciation of the Physical Presence of the Staff

Moos-Bjerre has also investigated the perceived importance among library users of the physical presence of the staff. In general, 90% of the Danish library users do appreciate the physical presence of staff members to help and to guide (Moos-Bjerre & Moos-Bjerre, 2014, p. 92). Neither, here, the differences between the members of the two segments are particularly sensational. In fact, 95% of members of the segment "mature from the lower middle class" and 92% of the "cultural super users" share a significant appreciation of the physical presence of staff.

Again, it is important to notice how you ask. If you, instead of asking whether the respondent appreciate the physical presence of the staff in the library, ask whether more library staff would make the given user increase their library use, you will get quite a different result. Indeed, only very few, 6%, of the respondents here agreed that additional staff would make them use the library more (Moos-Bjerre & Moos-Bjerre, 2014, p. 92).

Prolonged Opening Hours and Increased Library Usage

Significant differences occur, however, when we compare the attitudes of the two segments on what changes in the library that are likely to increase their library usage. Whereas the two segments more or less agree on flexible physical access as a primary library interest, the answers of the two segments vary significantly when asked whether prolonged opening hours are likely to increase their library usage.

Only 2% of the "mature middle class" members indicate that prolonged opening hours would increase their library usage, whereas 24% of the "cultural super users" agreed on the same question. Overall, it is, however, mostly young people on higher education that are most likely to let their library usage influence by prolonged opening hours. In fact, 34% of this segment agreed with that (Moos-Bjerre & Moos-Bjerre, 2014, p. 88).

Cultural Policy Issues

On the whole, the "mature middle class" is a particularly difficult segment to influence. Indeed, 40% of the segment, for example, answered that they did not know what would make them use the library more (Moos-Bjerre & Moos-Bjerre, 2014, p. 169). Maybe, it is also more a cultural policy issue whether the library should concentrate on serving their most loyal user segments like the "cultural super users" or try to appeal to less motivated segments like the "mature middle class" users. Moos-Bjerre emphasizes that it is very important not to forget or displace the loyal core user segment to attract other segments.

We know from the Danish representative Moos-Bjerre (2014) survey that 27% of the members of the "cultural super user" segment have answered that prolonged opening hours would increase their library usage in the future (Moos-Bjerre & Moos-Bjerre, 2014, p. 165). Besides, because more books and increased availability of e-books together with prolonged opening hours constitute the top three of the segments preferred library usage increasing elements, it would probably be advisable to follow that track. Especially, if the options are more extra, but unstaffed, opening hours versus fewer, but staffed, hours added. We will now focus on two particular sociodemographic factors: age and gender.

Age, Gender, and Ethnic Minorities

Although the check-in data as mentioned do not allow sophisticated data analyses, some interesting patterns can nevertheless be derived. Through the software company Cordura that recently (in August 2016) has merged with the company, Bibliotheca, I have got access to data from a number of Danish libraries in different districts through the last 3–5 years. I have chosen two of them, Langeland and Jægersborg, for further examinations.

Langeland

Langeland is a predominantly rural district situated on an island with 12,861 inhabitants in 2013.

Fig. 4.1 illustrates the distribution of visitors by gender during the last three and half years from January 1, 2013 to August 2016.

The figure, clearly, shows that 56% of the open library visitors during the three and a half years were men. The total number of visits is 111,905. This male majority is interesting because the majority of users in Danish library surveys are usually women. In Langeland, however, only in 1 month out of 44, October 2013, a majority of women occurred.

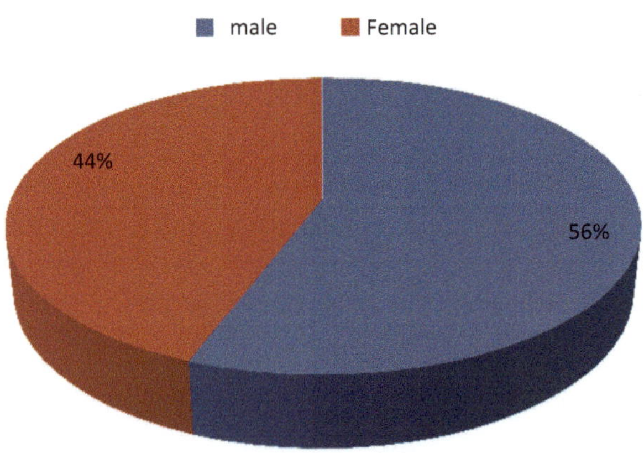

Figure 4.1 Langeland 2013–2016—visitors broken down by gender $N=111{,}905$. Cordura (2016)

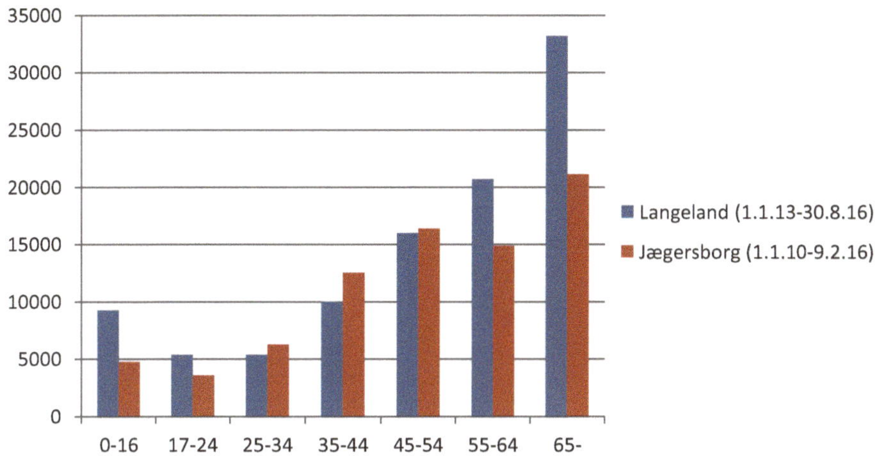

Figure 4.2 Visitors by age—Langeland and Jægersborg.
Cordura (2016)

In Jægersborg, a suburban district near Copenhagen, we find the same pattern. In the 5 years between January 2010 and August 2016, the open library had 80,187 visits of whom 56% were men. However, other Danish open libraries, for example, in Brande and Ikast, do not exhibit the same male majority.

Fig. 4.2 above shows the total number of visits in the two open libraries in Langeland and Jægersborg during a, respectively, 3 years and 8 months and 5 years and 1-month period. In both libraries, the number of visitors is increasing the older the visitors are, reaching its maximum at the age plus 65. Thus, Fig. 4.2 does not document that young fathers make up a numerically dominant user segment in the open libraries in Langeland and Jægersborg. On the other hand, the figures still allow a considerable number of young fathers with children to have been present, for example, during weekends in the staff-less opening hours of Jægersborg library from 2010 to 2016.

However, Fig. 4.2, certainly, gives the impression of the typical open library user as an elderly or old person. In Langeland, a consistent pattern reveals that from 35 years old, increased age also means increased presence in the library during the unstaffed opening hours.

Open Libraries and Ethnic Minorities

Evidence from a recent user survey in Copenhagen conducted in the spring 2016 adds some interesting details. The Copenhagen survey that is also referred to in Box 4.1 confirms the predominance of men, known from both Langeland and Jægersborg. Besides, the Copenhagen 2016 survey reveals that people with a non-Danish ethnic origin outnumbers the ethnic Danes during the staff-less opening hours. This observation is also worth mentioning because the public libraries in Denmark had played an important role as the most used public institution among the ethnic minorities.

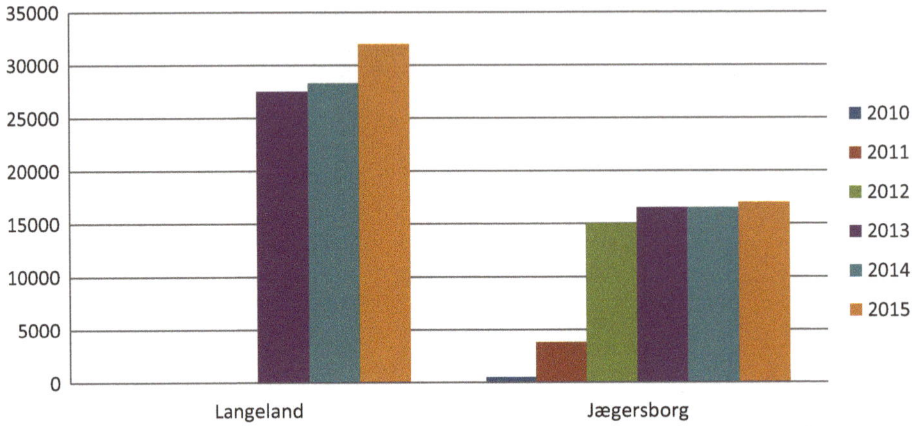

Figure 4.3 Visitors per year 2000–2015 Langeland and Jægersborg.
Cordura (2016)

It appears as if the open libraries also attract these user segments. It may have something to do with the open library's image as a "low-intensive" meeting place (Audunson, 2005). Finally, the Copenhagen 2016 survey documents that people with children make up the second largest user segment in the open libraries in Copenhagen. For natural reasons, the number of accompanying children are probably underestimated in the figures from Langeland and Jægersborg. Because small children do not, usually, possess ID cards, they are not counted when they enter the library together with their parents, grandparents, or siblings.

We will finally take a look at the development of the number of visitors during the most recent years.

Fig. 4.3 above illustrates two different development patterns. Langeland is a late starter but reaches a high level fast, whereas the early starter, Jægersborg, spends 2 years to reach a stable, relatively high level of visitors. The two patterns probably reflect that open libraries in 2010 in Denmark were still an unusual library service, whereas the staff-less model in 2013 when Langeland started had become a widespread and well-known library service. We will now look at the users and uses of the open libraries in other countries.

National and International Variations

Here the sources are much less detailed in many respects. However, comprehensive surveys have been made both in Finland and Sweden. These surveys have already been referred to in Chapter 3, in the section of Nordic experiences. Table 4.1 shows the number of logins and the number of open library units.

Table 4.1 illustrates two things. First, the eye-catching difference between Denmark and the other two Scandinavian countries. Second, the impressive growth rates as regards logins. In Norway, where the number of units grew about 10 times the logins

Table 4.1 Cordura Open Libraries and Logins in Denmark, Norway and Sweden 2014–2015

	2014		2015	
Country	**Number**	**Logins**	**Number**	**Logins**
Denmark	79	600,875	94	830,938
Norway	1	644	10	18,802
Sweden	3	953	7	4,730

Cordura (2016).

increased by a factor 29 and in Sweden where the number of units more than doubled, the logins grew to a size five times as high.

What Lies Behind the National and International Variations?

By national variations we refer to local, regional, or sociological variations within a certain country, whereas the international variations refer to variations among different countries, parts of the world, or cultures. As we have also seen, such differences can be mostly historical, bound to specific contexts. For example, the difference between the emergence of staff-less libraries in Far East Asia, where they were first established in densely populated urban districts, and in Denmark and Sweden, where the first more-opened libraries appeared in rural environments is a good example of a historic difference. Today, more-opened libraries are found in both rural and urban districts in Sweden and Denmark. We have apparently passed the era where an open library in Western Europe was considered to be a specialty only for rural and suburban neighborhoods.

Surveillance and Closed-Circuit Television

Another conspicuous difference concerns the use of closed-circuit television (CCTV) and video surveillance. In Denmark and in other countries such electronic equipment was together with panoramic overview and absence of dark and messy corners in the physical library considered to be essential accessories in an open library. In Sweden, however, the use of such equipment was restricted by the law; therefore, most Swedish open library users avoid being monitored by cameras. Here the difference is rooted in different legal frameworks in the two countries.

Open Questions

However, other international variations are more difficult to explain. The Nordic countries, Denmark, Finland, Norway, and Sweden are all relatively rich democracies. And all of them have introduced the staff-less library model and today they all have a number of open libraries in operation. In contrast, we have Germany, also a rich democracy with a well-educated population. But in Germany with its 82 million inhabitants there is only one staff-less library around the city of Hamburg. Are open libraries a specialty of small states like Denmark, Finland, Norway, and Sweden? Probably, but the model also seems to work in really populous countries like China and Japan. Another probable explanation refers to more complex cultural dimensions such as social capital.

Trust and Social Capital

Sociologists have pointed out that the measured level of trust and social capital exhibit significant differences between the Nordic countries on the one hand and Germany on the other. According to recent World Value Surveys (WVS), so-called "historically protestant" countries like, for example, Denmark, Finland, Norway, and Sweden, and Confucian countries like China, Japan, Taiwan, have a high level of trust in common. But then what about Germany as well as the United Kingdom and the United States that also belong to the category "historically protestant" countries?

Population Density

It is interesting that open libraries have both been used to protect thinly populated rural communities and to support quite modern 24/7 commuter life styles. Today, open libraries are found in many different contexts. Maybe, it is too early to determine why the open library model apparently thrives more in some contexts than in others. We have already noticed that the alleged differences between town and country showed up to be less important than believed, and, maybe, we need more time before we can determine fairly safe where the open library models fit and where it does not fit. We will now look at what we know about the open library users.

User Types and Segments

According to the nature of things, we do not have much information on the staff-less library users. The electronic access machines, certainly, collect some basic information about the visiting user such as their gender and age, but according to privacy regulations such data are not kept for a very long time, except when, for example, vandalism, or other criminal actions have taken place. Moreover, such data are, usually, not publicly available but reserved for the library staff. In cases of violations and crimes, the police has, in many countries, access to the CCTV files too. However, some of the new trends, for example, the transformative library (see Chapter 3), require much more detailed information about the visiting users than just their gender and age.

I was so lucky, when I prepared my first report on open libraries in Denmark (Johannsen, 2012a, 2012b), that the open library software supplier, Cordura, which stood behind the world's first open library in Gjern, made available for my research, data from eight Danish open libraries. It should be mentioned, here, that all the eight public libraries, in advance, had given their written permission to Cordura to deliver their data further to a third party representative. For this book, I have also obtained access to more current user data through Cordura and Bibliotheca.

The more recent user data were interesting, although they also left a lot of interesting questions unanswered. They did, for example, not allow comparisons between user behavior during the staffed and the unstaffed opening hours because identity data on visitors were usually not collected during the staffed hours.

Danish 2011 Cordura Figures

Despite such limitations, the 2011 Cordura data reveal a number of interesting, but hardly surprising insights about the users of the eight open libraries.

- gender,
- age, and
- preferred access time

The distribution by gender showed that 59% of the selected Danish users in 2011 were women and 41% men. This is interesting because newer data show a significant predominance of male users. A majority of women are seen in many public library surveys, so the widespread myth of open libraries as a haven for young men, being tired of overzealous librarians could not be verified. The female majority pattern recurred in all the participating libraries, except one, where the male sex constituted 57%. In recent Cordura data, this female dominance, however, has apparently been changed a lot.

There have been a discussion, at least in academic library contexts, about the relationship between gender and unstaffed libraries.

As to age, the 2011 Cordura figures show that in Table 4.2.

Again, more recent data from Cordura reveal a somewhat but not fundamentally different distributions as to age.

Economically Active Users

Table 4.2 shows that 71% of the open library visitors are within the economically active age. This is an interesting observation because one of the purposes of the

Table 4.2 Open Library Users, Denmark 2011 Age, *n* = 49.000 Open Library Visits

Age (Years)	Percent
7–16	12
17–34	18
35–44	21
45–64	32
65–	16
Total	100

Cordura (2011).

prolonged opening hours, exactly, has been to make it possible for busy people and commuters to benefit from the local library too. Also, the fact that about a third of the visitors are below 35 is worth noting at a time where the children's library are losing ground.

Few Early Birds and Few Night Hawks

In contrast to gender and age, the data about preferred times for open library visits seem also to deliver some practical usable, but not particularly surprising, information. The 2011 Cordura data, here, indicate that early mornings and late evenings do not seem to be particularly popular in the selected eight libraries. The open library rush hours are located in the periods from 9 a.m. to 12 a.m. and from 3 p.m. to 8 p.m. Such preferences, however, could be different in more urban, creative class environments. Nevertheless, 24/7 preferences do not seem to have had their breakthrough in common provincial towns and rural surroundings yet. Because many open libraries are situated in rural areas to counteract stagnation and depopulation, the fulfillment of the 24/7 vision does not necessarily need to be among the most urgent priorities. However, recent Swedish interview data indicate that weekend evening opening hours are popular among commuters.

Some American Academic Library Users Have Different Preferences

Not only members of the urban, creative class might have other preferences as to opening hours than the average Dane. Students have too. A survey (Lawrence & Weber, 2012) of American academic library student users reveal that patrons place high value on late-night access to the library. In general, they view the library as a quiet, peaceful place conductive to studies. The survey also shows that opening hours extended to 2 a.m. Sunday through Thursday was very popular with the students. This represents an example where academic library users have quite dissimilar expectations when compared to users of public libraries.

Other Danish User Surveys

In addition to the Cordura data, a small, four out of the 81 open libraries in 2011, number of Danish staff-less libraries had collected data on users and use in 2011. Among them, a questionnaire-based survey conducted in April 2011 in a small provincial town, Dronninglund, could be interesting (Johannsen, 2012a). Dronninglund has about 3500 inhabitants. As to the basic variables, gender and age, 72% of the respondents were women and 70% plus 40 years. Of the users of the open library, 64% had been users for more than 5 years, thereof 36% in more than 20 years. Only 14% were new users with less than 1 year's library usage. The survey also showed that 70% were using the open library to deliver or pick up ordered materials; 30% primary purpose was to read newspapers and journals.

Increased Library Usage

Most interesting, however, was the result that 38% indicated that their use of the library had increased because of the open library. Also, worth noting is that 18% mentions the meeting place function as important for their use of the open library. A majority points at the increased flexibility, due to longer opening hours, as the main attraction of the staff-less library, whereas few, 8%, point at the calm and space to yourself as the main benefit. Figures from the other three small 2011 Danish user surveys, more or less, correspond to the Dronninglund findings.

To sum up, the results of the earliest open library user surveys in Denmark, after about 1 or 2 years experience with staff-less libraries, mostly not showing worth mentioning differences as to gender and age, compared to the user's using the library during the staffed hours. Moreover, the data indicate that the presence of the staff-less opening hours rather seems to increase library use, in general, than the opposite. However, the Cordura data and the data from early user surveys leave a number of vital questions open. Fortunately, a recent, representative survey, *Fremtidens Biblioteker* [*The libraries of the future*] (Moos-Bjerre & Moos-Bjerre, 2014) offers a number of new insights as to open library users.

A Representative Danish User Survey—The Libraries of the Future 2014 Report

The survey was initiated by the Danish Library Association (Danmarks Biblioteksforening). Data were collected in 2013 through a representative, questionnaire-based survey sent out to 2000 library users and nonusers the age 15 years or older. Besides, supplementary qualitative studies (focus group interviews, in depth interviews, workshops) were conducted.

The survey that has already been mentioned is interesting for several reasons. First, because it tries to uncover how Danish users appreciate and use the open libraries. Furthermore, however, it tries to specify and nuance our knowledge about which type of users who especially like and use the open libraries. Second, *The Libraries of the Future* is based on advanced, statistical cluster analyses, which enable the construction of multifaceted user categories based on both demographic, library use patterns, and life-style characteristics.

Books—Still a Primary User Interest in 2013

Among the primary user interests concerning public libraries, the traditional opportunity to lend physical books is number one; 77% of the respondents, when asked about their primary interest in using the library, mention this purpose. In the questionnaire, the respondents could mention three primary library interests. Compared to book lending, activities like lending music on CDs and films on DVD and lending e-books were chosen by, respectively, 26% and 15% of the respondents.

Flexible Physical Access—Appreciated by Many Users

In our context, the most interesting result is, however, that 29%, pointed at flexible, physical access, and self-service as one of their primary interests in public libraries. Hereby, the open library option took the second place among the main library preferences of the Danish users (Moos-Bjerre & Moos-Bjerre, 2014, p. 80). Both the importance and the positive assessment of staff-less libraries, hereby, seem to have been confirmed by a reliable, representative survey.

Prolonged Opening Hours Will Increase Library Usage

But *The Libraries of the Future* contains further interesting information about user preferences as to open libraries. When asked about their future library use, 30% of the users indicate that they would use the library more if the opening hours were prolonged, and, which is important, they would also accept to achieve the longer opening hours, despite a part of them being staff-less (Moos-Bjerre, 2014, pp. 87–97).

Which User Segments Appreciate Open Libraries Most?

But who among the many different types of users were specially in favor of longer, partially staff-less opening hours and who were the least excited users? Although the members of the 10 constructed user segments, in general, shared the opinion that flexible opening hours were very important, the points of view as to whether prolonged opening hours would lead to increased library use differ a lot. Especially, young (20–29 years) people following a higher education and the segment of so-called "cultural super-user," typically, middle aged, fiction-reading women, would increase their library use if the opening hours were prolonged. During the focus group interviews, the participants were asked whether they preferred extended opening hours without staff or more staff, almost all preferred the longer opening hours (Moos-Bjerre & Moos-Bjerre, 2014, p. 88). A female cultural superuser explains:

> *Well, long opening hours because I am, pretty much, doing myself, I not so often need assistance from staff. Usually, I order through the Net and the pick up the materials. I seldom need help*
>
> *Moos-Bjerre (2014, p. 88)—translated by CGJ.*

The Least Excited Users

Among the least excited users, one sees, elderly members of the lower middle class, young parents (from 20 to 29), and seniors (typically, an older, retired man). These figures, to a certain degree, seem to support, the point of view of some critics, claiming that the socially weaker library users are the least interested segment concerning the staff-less library model.

However, a majority of Danish library users appreciate the open library model and this does not mean that they are indifferent to staffing. On the contrary, 90% of all users find it important that the staff is physically available in the library to help and supervise (Moos-Bjerre & Moos-Bjerre, 2014, p. 92). In contrast to the positive assessments, where 30% of the users indicate that they would use the library more with prolonged openings hours, only 6% specify that more library staff would make them use the library more (Moos-Bjerre, 2014, p. 92).

Usage Patterns and Amount

We have already dealt with some aspects of usage: preferences as to library visit times. In this section, we will focus two further aspects of library usage:

- Visits/logins
- Loans

Visits/logins

Because there are available data, which enables us to compare visits and loans during both the staffed and unstaffed opening hours, at least in Denmark, the figures make interesting analyses possible. At the same time, it should be mentioned that the data concerning all Danish staff-less libraries are rather incomplete because data from several libraries are missing.

Differences Between the Counting of Visits in the Staffed and the Unstaffed Time

Indeed, when I prepared my survey of staff-less libraries in Denmark, there were only comparable data about 34 of the 81 participating libraries (Johannsen, 2012a). It should also be mentioned that the sources of visits in the staffed and unstaffed periods are quite different. In the unstaffed opening hours, the primary source is data from the check in procedure. This procedure, however, often includes a risk of underestimating the number of visitors. Because the check-in requires the use of an identity or health card, accompanying children, family members, and other users often slip into the library together with the card holder. In the staffed opening hours, ID cards are not used. Instead, some libraries have installed mechanical counting apparatus at the entrance.

As it is more difficult to let more persons slip through such counting apparatus at the same time. Therefore, are the data from this source more reliable than data from ID card check in procedures. In practice, it means that you, probably, should time the unstaffed in locks with two or three to obtain realistic figures.

Significant Differences—Why?

Table 4.3 shows the number of visits per hour in 2011 in 10 public libraries arranged after number of visits per hour during the unstaffed opening hours.

Table 4.3 Staff-Less Libraries, Denmark, Visits per Hours During Unstaffed and Staffed Opening Hours, 2011, $N=34$

	Visits per Hour Unstaffed	Visits per Hour Staffed	Name of Library
1	16.6	33.6	Hornbæk library
2	11.2	43.0	Jægersborg library
3	7.9	63.9	Skødstrup kombi library
4	5.1	11.0	Christiansfeld library
5	5.1	23.0	Langaa library
6	5.1	16.9	Ganløse library
7	5.0	9.5	Vamdrup library
8	4.8	15.8	Skævinge library
9	4.1	23.2	Graasten library
10	4.1	36.9	Løgstør library

Johannsen, C. G. (2012a). *Åbne biblioteker – et kortlægningsprojekt [Open libraries – a mapping project]*. Copenhagen: Det Informationsvidenskabelige Akademi.

The data from the two different sources, however, showed that the number of visits during the staffed hours were always bigger than during the unstaffed hours. Considering the mentioned, estimated understatement as to the number of unstaffed visits, due to the different access procedures, the status of the unstaffed opening hours will be radically changed in several cases. If we, for example, time the unstaffed visits by two, one library (number 7) will achieve more visits during the unstaffed than during the staffed opening hours. If we, however, assume that the figures should be timed by three, two more libraries (number 1 and number 4) could be added.

There seems to be many explanations and reservations. First, because most of the open libraries in Table 4.1 had only been operational in a short time, about 1 year, one might think that as soon as the users become familiar with the open library offer and its opening hours, the staffed and unstaffed figures would probably approach each other further. Evidence concerning visits in five units of public libraries in Sønderborg from 2010 to 2014 indicates that this assumption is right. In fact, the number of visits during the unstaffed opening hours has increased from about 40,000 visits in 2011 to about 50,000 in 2014 (Biblioteket Sønderborg, 2016, p. 22). Second, you can also explain the significant difference between the staffed and unstaffed opening hours, with the conclusion, that the libraries, apparently, have been successful in hitting the spot as to the users' preferences concerning opening hours. Finally, there could exist a certain X-factor or feature, associated with a staffed public library, that was able to explain why the staffed opening hours were much more popular than the staff-less.

A similar pattern is revealed in Fig. 4.4. The figure shows the number of unstaffed logins and staffed visits from 2010 to 2015 in six Danish open libraries.

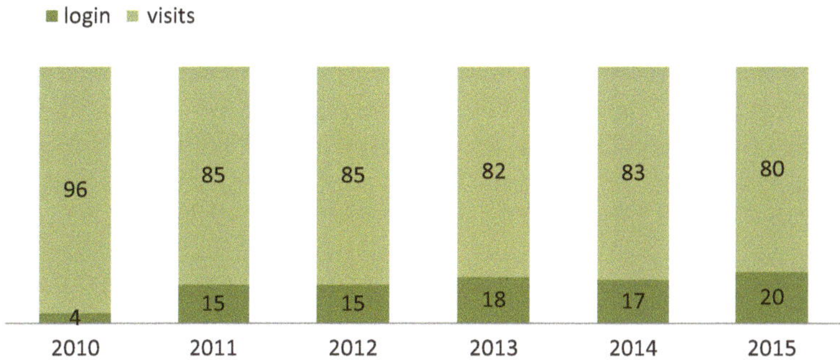

Figure 4.4 Logins (unstaffed) and visit (staffed) 2010–2015 in six Danish open libraries percentages. Note: The six libraries in Sønderborg municipality were Broager, Dybbøl, Gråsten, Hørup, Vester Sottrup, and Nordborg—the total number of logins and visits in 2010 were 247,578 and in 2015 312,347.
Sønderborg libraries, Denmark.

The number of visitors appear to be significantly lower during the unstaffed than during the staffed period. In 2010, for example, there were about 10,000 logins and 237,000 visitors, whereas in 2015 there were 62,000 logins and 250,000 visitors. If we, however, assume that the number of logins underestimate by two or three, there might be between 120,000 and 186,000 visitors behind the 62,000 logins. The visitors during the staffed hours still form a majority but the distance between the columns is probably lesser than Fig. 4.4.

"Use the Library When It Fits Me Best"—the Randers Survey (2012)

In the Danish town, Randers, an interesting user survey, based upon questionnaires and 374 respondents, was conducted in March 2012 (Lerche, 2012a, 2012b); 81.8% of the respondents indicate that they use the library during the unstaffed hours, whereas 16.8% do not use the library in the unstaffed hours. Interesting is also that 89.8% use the library in the staffed hours; 9% say that they do not use the library during the staffed hours; 96.6% indicate that it is easy for them to loans and deliver books themselves. A typical loaner statement sounds:

> *I'm very content with the opportunity to use the library when it fits me best. If I need help, I try to accommodate my visits till when staff is available, but otherwise I lock myself in when it suits me. Before I only used the library four times year, but now I come there several times a week*
>
> *Translated by CGJ.*

That the final sentence is not exceptional appears from the fact that 31% of the respondents indicate that they had become new users after the introduction of the open library concept in Randers.

Norwegian Figures—Brukeradfærd på storbybibliotekerne *(2015)*

In Norway, a recent survey (Brukeradfærd på storbybibliotekerne, 2015) revealed some interesting trends. The survey covered user behavior in city libraries in Norway (Oslo, Bergen, Stavanger, Kristianssand, Trondheim, and Tromsø). One of the findings was that the number of users who addresses the staff during a library visit was 14%. From 2007 to 2015 this number had decreased from 25%. On the other hand, the percentage of users who did not make use of the services of the library during all or most of their time during the visit grew from 9% to 16% from 2007 to 2015. This development reveals the growing of the library as a meeting place.

Besides, the data concerning the six city libraries, the survey contains a comparative study of library user behavior in two Oslo branches, Majorstuen and Lambertsseter, where unstaffed opening hours together with staffed opening hours had been introduced. Here, it is possible to compare certain aspects of user behavior in the staffed hours with behavior in the staff-less.

More Study and Work During the Unstaffed Hours

As to the basic public library activities like loans and deliveries, the user behavior is approximately the same during the staffed and unstaffed hours. However, in both branches, significantly more users study and work during the unstaffed than during the staffed hours. In Majorstuen, 19% works during the staffed and 28% during the unstaffed and the corresponding numbers for the Lambertseter were 17% and 31%. However, the survey does not examine the underlying causes of this difference. Is the difference due to a more quiet atmosphere during the staff-less hours or is it because the staff-less opening hours are more convenient for the students?

On the other hand, it is obvious that traditional public library activities like loans and deliveries are more common user activities in the two branches during the staffed than during the unstaffed periods. A total of 47% and 57% loan or deliver library materials during the staffed and 42% and 46% during the unstaffed. These figures point in the same direction as the Danish, where the loans are much higher during the staffed opening hours than during the unstaffed hours. The percentages of users who make little use of the library services during their visits are approximately the same: 14% and 16% during the staffed and 16% and 18% during the unstaffed hours. These figures could, with cautious, be interpreted as an indication of the staff-less library being more seen as a working place than as a meeting place by its users.

Spending More Time in the Open Library

Box 4.3 illustrate that the open library model influences the time users spend in the library.

Loans

Regarding loans, the source conditions are less complex because it is the same automation system that handles the loans and deliveries during both the staffed and staff-less opening hours. On the other hand, even fewer Danish libraries had calculated the loans

> **Box 4.3 Gladsaxe Libraries (Denmark)—Case**
>
> *Results since introducing open + to its residents Gladsaxe Libraries have seen an increase in both the number of people that are visiting the library and the facilities that are being used; now that it is open for longer, community members are not just coming to the library to pick up and return their items, they are spending more time utilizing the different services that the PC, or they may not have so much space at home, so it is difficult for them to research and study, now with the extended access they can come here until whenever they want, it is a real haven for them.*
>
> Open+ brochure, issued by Bibliotheca (2016).

for the staffed and staff-less opening hours separately. In my 2011 survey, only 21 out of 81 Danish public libraries with staff-less opening hours had reported the loan data in a way that could enable comparisons between staffed and unstaffed opening hours.

The same trend as seen by the visits showed up here too. In average, there was also here a significant difference in 2011 between the number of loans in the staffed and in the unstaffed hours. During the unstaffed hours, the average number of loans per hour was four, whereas the corresponding figure concerning the staffed hours was 29 (Johannsen, 2012a, p. 24). The calculation of the average numbers is based on all the 21 libraries, not only on the 10 libraries listed below. The figures are also influenced by the circumstance that the staff-less services were started in different months during the year (Table 4.4).

Table 4.4 Staff-Less Libraries, Denmark, Loans per Hour and Total, During Unstaffed and Staffed Opening Hours, 2010/2011—Top 10, $N = 10$

	Loans – Staff-Less per Hour	Loans – Staff-Less Total –1000	Loans – Staffed per Hour	Loans-Staffed Total –1000	Library (Staffed/ Staff-Less Hours per Week)
1	17.3	30	20.9	21	Tørring (28/48)
2	8.9	26	20.9	6	Christiansfeld (8/80)
3	7.7	22	17.8	5	Vamdrup (8/80)
4	5.5	3	15.3	5	Jægersborg (37/59)
5	5.5	16	16.3	5	Lunderskov (8/80)
6	5.3	14	19.7	11	Langå (16/75)
7	4.9	9	39.0	26	Fensmark (21/56)
8	4.7	15	32.4	26	Dronninglund (25/98)
9	4.4	12	9.9	4	Tjæreborg (15/98)
10	3.5	5	52.5	24	Aars (28/98)

Johannsen, C. G. (2012a). *Åbne biblioteker – et kortlægningsprojekt [Open libraries – a mapping project]*. Copenhagen: Det Informationsvidenskabelige Akademi, Table 3.

On the other hand, it is an eye-catching feature that the total number of loans in 6 of 10 libraries is significantly higher during the unstaffed than during the staffed opening hours. In, for example, library number 3 (Vamdrup) and 5 (Lunderskov), the loans during the unstaffed hours were three to four times higher than in the staffed. However, the library in Vamdrup is unstaffed 80 h per week and only staffed in eight. This relatively large number of unstaffed opening hours explains why there were no library in Denmark in 2011, which loans more per hour during the staff-less than during the staffed opening hours. In business terms you may say that the production (of loans) was often higher during the staff-less opening hours than during the staffed, whereas the productivity (in terms of loans per hour) in 2011 in Denmark was always higher in the staffed opening hours.

Also in Sweden, the introduction of open libraries has led to increased loan figures, which the following quote by library director, Eva Bünger, Ale municipality, illustrates:

> *The branch libraries were threatened, but today nobody dare to touch them. Since we installed "More Opened" our loans have increased with 17 percent in Surte and 12 percent in Skepplanda.*
>
> *Translated by CGJ—Axiell. http://www.axiell.se/produkter-tjanster/for-lantagare/axiell-meroppet/.*

Fig. 4.5 illustrates the development of visits (=logins) and loans during the unstaffed hours at a Danish library from 2013 to 2015. It is interesting to see how the loans from 2010 to 2011 grow much faster than the visits after which they move more evenly. It difficult to say what precisely has caused this movement.

Importance of a Proper Balance Between Staffed and Unstaffed?

A nearby hypothesis that, maybe, could explain some of the variations could be that the balance between staffed and staff-less opening hours is crucial. It could, for example, mean that libraries with a more balanced distribution around the average, for example,

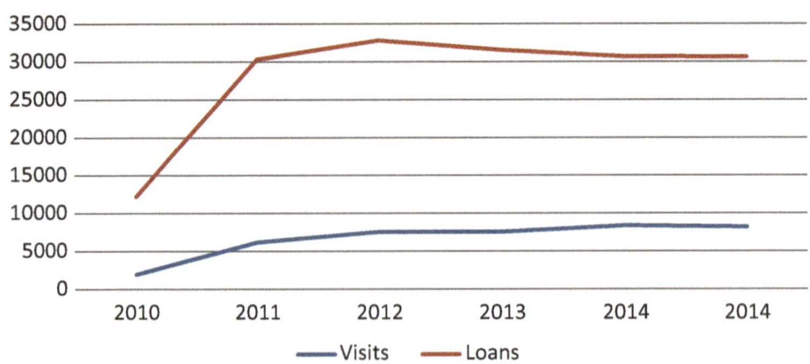

Figure 4.5 Dybbøl library 2013–2015 logins and loans during unstaffed hours. Bibliotheca (2016)

20 weekly hours staffed and 80h unstaffed would be likely to perform different than libraries with a more skewed distributions, especially, with relatively few staffed hours. This interesting question is examined in Table 4.5 below. The 10 chosen libraries are selected from both the top and the bottom concerning visits per hour (Table 4.5).

To get an overall impression of the figures, the staffed and unstaffed hours of the two groups, number 1–5 and 6–10, are added up, resulting in the figures below:

a. 113/258 (staffed/unstaffed), highest number of logins
b. 73/335 (staffed/unstaffed), lowest number of logins

It seems as if there is a connection between the percentage of staffed opening hours of the total number of opening hours and the number of logins. Category "a" contains libraries with 30% staffed hours out of the total opening hours, whereas "b" contains the five libraries with the lowest numbers of logins. The latter category, b, is also characterized as the category with the lowest percentage of weekly, staffed opening hours, 18%. Both categories, however, also contain exceptions. As the tables above are based on relatively few units, it does not seem to be possible to make further generalization as to what is the ideal distribution between staffed and staff-less, weekly openings hours with the purpose of achieving more logins or loans.

Most important is, probably, special local contexts and specific demographic circumstances. Otherwise, it is difficult to explain why the library in Harlev with a staffed/unstaffed distribution 15/26 has a much lower number of visits per hour than Skødstrup library with a similar distribution and Ganløse with an even more skewed distribution. The table, however, tells, which is also important, that libraries with a reasonable amount of opening hours, also seem to attract users during the staff-less hours.

Table 4.5 Staff-Less Libraries, Denmark, Logins per Unstaffed Hour and Balance Between Staffed and Unstaffed Opening Hours, 2011 – $N = 10$

	Logins per Hour Unstaffed	Balance Between Staffed/Unstaffed Hours per Week	Library
1	16.6	39/38	Hornbæk
2	11.2	37/59	Jægersborg
3	7.9	15/15	Skødstrup
4	5.1	14/66	Ganløse
5	3.7	8/80	Lunderskov
6	1.5	8/88	Vester Sottrup
7	1.4	8/88	Dybbøl
8	1.3	21/56	Fensmark
9	0.7	21/77	Allinge
10	0.1	15/26	Harlev

Johannsen, C. G. (2012a). *Åbne biblioteker – et kortlægningsprojekt [Open libraries – a mapping project]*. Copenhagen: Det Informationsvidenskabelige Akademi, Table 5.

Table 4.6 Staff-Less Libraries, Denmark, Loans per Hour and Balance Between Staffed and Unstaffed Opening Hours, 2011, $N=7$

	Loan per Hour Unstaffed	Balance Between Staffed/Unstaffed Hours per Week	Library
1	17.3	28/48	Tørring
2	8.9	8/80	Christiansfeld
3	7.7	8/80	Vamdrup
4	5.5	37/59	Jægersborg
5	2.4	28/98	Farsø
6	2.2	28/98	Løgstør
7	1.3	24/74	Åkirkeby

Johannsen, C. G. (2012a). *Åbne biblioteker – et kortlægningsprojekt [Open libraries – a mapping project]*. Copenhagen: Det Informationsvidenskabelige Akademi.

The same picture is seen in Table 4.6, when considering the spread of the number of loans per hour during the unstaffed time and the distribution between staffed and unstaffed opening hours (Table 4.6).

Here, we also find both evenly libraries (Tørring and Jægersborg) and unevenly distributed libraries among the high performing libraries.

Growth

Growth both in numbers of visits/logins and loans are often presented to proof the success of the open library concept. In Denmark, it has been emphasized that the number of visits and loans have demonstrated remarkable growth rates in times, where the number of traditional book loans usually stagnates or decreases. A figure that often has been mentioned in the public debate is 23%, which represents the average growth rate as to the total number of library visits (Johannsen, 2014b). The figure is without doubt impressive; however, it should be kept in mind that this growth rate is achieved by increasing the opening hours, not only by 23%, but often by many 100%. The average balance between staffed and unstaffed opening hours in Denmark is 20 staffed to 80 unstaffed hours per week (Table 4.7).

Because data on loans before and after the introduction of staff-less opening hours were only available from 25 out of the 81 Danish staff-less libraries in 2012, there is likely some uncertainty associated with the figures above. We do not know, for example, if the missing libraries, predominantly, represent libraries with lower or higher growth rates. Among the 25 libraries who gave up before and after data, 11 had perceived decreasing numbers of loans after they became staff-less. A further uncertainty derives from the fact that in a few cases, where some monthly

Table 4.7 Staff-Less Libraries, Denmark, Yearly Number of Loans (in Thousands)—Before and After Establishment of an Open Library, Libraries With Growth, Top 10, Only Libraries With More Than 2% Yearly Growth

Library	Start Year/ Month	Staffed/ Unstaffed Hours per Week	Loans Before (Year)	Loans After (Year)	Difference Growth Percentage (%)
Broager	2010/01	22/71	31(2009)	41(2011)	+32
Vester Sottrup	2009/09	17/76	23 (2009)	28(2011)	+24
Dybbøl	2010/01	17/76	24 (2009)	29(2011)	+22
Gram	2010/02	32/39	34(2009)	40 (2010)	+18
Dronninglund	2011/02	25/98	54(2010)	61(2011)	+13
Ågerup	2010/05	12/64	17(2009)	20 (2010)	+12
Jelling	2010/10	19/72	46(2009)	49(2010)	+7
Skævinge	2011/02	12/60	54(2010)	57 (2011)	+6
Hørup	2010/03	17/76	28(2009)	30(2010)	+6
Christiansfeld	2010/01	8/80	37(2009)	39(2010)	+6

Based on Table 7, Johannsen, C. G. (2012a). *Åbne biblioteker – et kortlægningsprojekt [Open libraries – a mapping project]*. Copenhagen: Det Informationsvidenskabelige Akademi.

loan data were missing, the absent data were calculated based on the data from the remaining months.

Table 4.8 represents the figures from the libraries with decreases more than 2% (Table 4.8).

Table 4.8 illustrates that the establishment of an open library solution is not always accompanied by increasing loan figures. The two libraries with the highest downturns. Ejstrupholm and Engesvang, are, however, both atypical public libraries with in advance small numbers of loans. In fact, one can wonder why the two libraries are able to preserve their loan numbers 70%–80% of their loans being full time staff less. It is difficult to point out why the other four libraries on the list have perceived reductions in the number of loans. In general, the loans, especially of books, have declined since the 1980s in Denmark. The balance between staffed and unstaffed, weekly opening hours could probably have played a role. On could assume, that libraries with a fair balance between staffed and unstaffed opening hours were more likely to achieve progressive loan figures. However, the data leave ambiguous interpretations. It is true that the average number of staffed opening hours is higher, 18, among the growth than among the libraries with waning number of loans. The average number of unstaffed opening hours is, however, also higher among the growth libraries, 71, than among the libraries with decreasing loans, 63. Besides, we find libraries with a very uneven distribution of staffed/

Table 4.8 Staff-Less Libraries, Denmark, Yearly Number of Loans (in Thousands), Before and After Establishment of an Open Library, Top 6 Libraries With Decreasing Number of Loans

Library	Start Year/ Month	Staffed/ Unstaffed Hours per Week	Loans Before (Year)	Loans After (Year)	Difference Reduction Percentage (%)
Ejstrupholm	2011/02	0/52	5(2010)	3(2011)	−34
Engesvang	2011/01	0/33	6(2010)	5(2011)	−17
Vamdrup	2010/01	8/80	39(2009)	34(2010)	−14
Sæby	2010/03	35/64	122(2009)	106(2010)	−13
Gudhjem	2010/06	15/83	18(2009)	16(2010)	−7
Børkop	2010/04	26/65	79(2009)	76(2010)	−3

Based on Table 7, Johannsen, C. G. (2012a). *Åbne biblioteker – et kortlægningsprojekt [Open libraries – a mapping project]*. Copenhagen: Det Informationsvidenskabelige Akademi.

unstaffed opening hours in both categories (Christiansfeld with a 6% growth rate and Vamdrup with 14% decline). Again, local and specific factors seem to play a significant role too.

User Purposes, Needs, and Preferences

It has been mentioned that the open library concept is very much dependent on advanced information and communication technology such as radio-frequency identification, CCTV, and automated delivery and loan systems. But modern information technology could also be regarded as a serious competitor to the open library concept. Why use time to visit a physical library if you can access the same information sources from your home address? Why spend time to pick up a novel at the local library when you can easily borrow an e-book through your home PC or tablet?

Why the Physical Library?

Here, the unique qualities of the physical library apparently play a significant role. Also, the opportunity to arrange meetings would, usually, be more convenient using the facilities of a public library than home. In academic library contexts, some believed that having resources available electronically 24/7 would enable libraries to reduce hours and depend upon an only virtual presence to satisfy the needs of their users. However, as academic libraries have realized that although electronic devices are popular, it does not replace the need for a physical space dedicated to group and individual study (Lawrence & Weber, 2012). The same is probably valid for public libraries, although for somewhat different reasons.

Academic Library Users

The primary reason for students coming to the library midnight to 2:00 a.m. was "quiet study", whereas number 2 was computer use. They found the distractions of home, room-mates, food, TV, the noise of neighbors' parties, at odds with the need to tackle academic assignments (Lawrence & Weber, 2012, pp. 536–537). The need for a quiet place is certainly also valid for a public library user, although she would not prefer the hours around midnight for her library visits. Indeed, hours after 10 p.m. have showed up to be the least popular opening hours for open libraries in Denmark.

User Behavior—and Misbehavior

A widespread fear among library professionals in several countries (e.g., Denmark, Germany, Sweden, the United Kingdom) has been and is still certainly that some users left to fend for themselves in unstaffed libraries would probably damage interior and expensive equipment during the staff-less opening hours. In Denmark, according to one of the staff-less library pioneers, this fear was prevailing in many libraries, planning to introduce the open library concept. In Germany, the same fear is probably the reason why a country like Germany only has one open library around Hamburg (interview Claudia Knauer in June 2016). Fortunately, the widespread fear, mostly, showed up to be groundless. A typical experience from the Danish Gladsaxe libraries sounds (Box 4.4).

Box 4.4 Typical Worries Before Installing an Open Library, Gladsaxe Libraries (Denmark), Case

" ... implementing an open library concept is a little like entering the unknown for a library, so it was only natural that Gladsaxe had some concerns." Maria Kruchov, Librarian and Branch Coordinator, comments: "At first we were worried about how patrons would take care of the library when staff are not present; we did not want to leave it in their hands only to come back and find that it is not in such good condition, but that has not been the case at all, there may have been a piece of litter left behind, but that's the worst of it."

Esben Fjord, Head of Development at Gladsaxe libraries, adds: "When we first started the project we were worried about people entering the library that were not registered with us and people misusing and damaging the facilities, however, that has not happened, everyone seems so happy that they have the opportunity to come to the library whenever they want that they have really taken responsibility of the library and its well-being.

One of the benefits of the open+ solution is that it incorporates CCTV cameras, allowing libraries to capture images and video footage to monitor each person that enters the library. Additionally, extra cameras can be placed around the library to provide extra peace of mind for both the library and its patrons.

Open+ brochure issued by Bibliotheca (2016).

Box 4.5 No Major Issues in Peterborough (United Kingdom)—Case

Since the initial implementation, over 1000 members have opted in, with a peak usage of 250 open+ users at Central Library in one day. Lisa Roberts was pleased to report no major issues: "The only slight problem we have encountered is accidental tailgating. Central Library is very popular and there are often queues of people waiting to get in. With open+ members having earlier access, patrons who had not yet opted in assumed that the library was open and wanted to follow them in. We noticed this during the early stages of the pilot, and used it as an opportunity to encourage more members to sign up.

Open+ brochure issued by Bibliotheca (2016).

The same worries were also found among the professionals in the United Kingdom and the same story about no major issues could be told there too. Notice that Peterborough in Eastern England is not a rural district (Box 4.5).

Instances of Vandalism in Denmark

A survey was conducted in 2013 by *Momentum*, the journal of the National Association of local authorities in Denmark, about vandalism, theft, and the use of libraries as warm shelters for homeless people (Jørgensen, 2013). Municipal managers within cultural and leisure time departments were asked about their experiences concerning staff-less libraries. Only municipalities with libraries were included in the survey. There were, besides "do not know", four response options: "to a large extent", "to a certain extent," "in a little scale", and "no". There were questions on vandalism, theft, use of the library as warm shelter, and citizens who could not handle the technical facilities (lock in procedures, automatic delivery and loan, and the like).

80% Had Not Experienced Vandalism

None of the responding managers had experienced problems "to a large extent". Four percent had experienced problems "to a certain extent" but only as to users with technical problems; 17% indicated that they had had problems with vandalism "in a little scale," 23% with theft "in a little scale," 25% with the library as a warm shelter "in a little scale," and, finally, 31% had experienced users with technical difficulties "in a little scale". It means that 80% had not experienced vandalism, 75% not theft, 70% had no warm shelter problems, and 56% experienced no users with technical problems. In addition, there were very few "do not knows."

Few Problematic Neighborhoods

Although vandalism, theft, and unrest appear to be a minor problem, instances have occurred in Danish provincial towns, for example, in Aalborg, and in the capital, Copenhagen, for example, around the Blaagaard branch library. Both the mentioned neighborhoods in Aalborg (Grønlands Torv) and in Copenhagen (Blaagaards Plads) have a reputation for gang crime and street violence. Chapter 5 contains an account of what has been done in Aalborg to counteract the unrest and it seems that most of the problems have been solved through a comprehensive approach including cooperation with local police and social service staff. It should also be mentioned that many urban districts both in Aalborg and Copenhagen and also, for example, in Norway's capital, Oslo, have not experienced serious incidences of crime, vandalism, and unrest. On the other hand, vandalism and unrest have not been restricted to urban environments. In Denmark, the open library in idyllic Præstø, situated in Southern Zealand by the sea, has been plagued by young trouble makers (Holmgaard Larsen, 2011a, 2011b; Box 4.6).

Vandalism in Sweden

A Swedish survey covering nine staff-less libraries in Skåne (Johansson et al., 2015) where the respondents were asked to what extent they agreed with the statement, that vandalism in the libraries had increased because staff-less opening hours were introduced in their libraries, and indicated that the absence of vandalism known from Denmark also characterized the situation in the Swedish Skåne province in Southern Sweden; 91.3% indicated that they did not agree with the statement at all, whereas 4.3% did agree partly. The group of respondents consisted of library directors and librarians from the nine libraries.

Vandalism in the United Kingdom

In general, vandalism is not mentioned as frequently occurring events in British open libraries. Like in the Scandinavian countries expectations sometimes have been unnecessary pessimistic. In Peterborough, in 2016, the Dogsthorpe branch library had suffered from antisocial behavior for a long time and was therefore chosen for the pilot test of the open library system. It showed up, however, that against expectation there were no problems associated with an open library in this particular branch library (Hitchcock, 2016).

American Experiences With Theft and Vandalism

Also in other countries than Denmark the absence of vandalism has characterized the introduction of the open library model. Although the Danes, together with the other Nordic peoples, get high scores as to trust in international surveys, absence of

Box 4.6 Sønderborg Libraries (Denmark)—Case

In 2008, the library's management put forward, to Sønderborg Councils' cultural affairs department, the idea of the establishment of three 'self-service' local libraries. This proposal was positively accepted. Moreover, the local politicians decided to raise the number of self-service local libraries to five, out of a total of eight local libraries. These five self-service libraries were established between 2009 and 2010. The libraries were now open from 8 a.m. until 10 p.m. on weekdays, and from 8 a.m. until 5 p.m. during weekends. The staffed opening hours remained the same as before. There were effectively no cut-backs. The new library services were seen by the politicians as giving added value. Furthermore, a more effective use of the already existing library resources were emphasized. About five years later, in 2014, a further library was set up with self-service functions. This was initiated politically, with regards to budget negotiations. Extra funds were made available for the library buildings and their operative costs.

There was, however, slight skepticism on the part of the staff, when they received the news that the libraries opening hours would be extended without staff being present. They anticipated disorder and vandalism. Luckily this did not occur. On the contrary, library users tidied up after themselves and each other. There were only a few episodes where library staff were met with disorder when they arrived at work. Indeed, the library users developed a strengthened sense of ownership and looked after 'their local' libraries. This matched Sønderborg's councils overriding strategy of 'active civic participation'.

The enthusiasm of the local press and citizens, about the extended services, caught on with the employees and turned their initial worries and skepticism into a sense of pride.

With over eight years of experience with self-service libraries, the conclusion is, that is has been a great success.

Citizens can use the services of the libraries when it suits them best. They can borrow and return materials, read newspapers and magazines, make photocopies, hold meetings, study and play. The solution Sønderborg chose, to retain many hours with professional staff available, continues to offer the opportunity to receive qualified help when needed.

What do the figures show? The number of materials borrowed from the libraries is statistically stable, however, the number of visitors to the libraries has increased dramatically. It is apparent that relational and personal communication is an important and crucial factor.

The majority of citizens prefer to visit their local library during staffed opening hours. It is our viewpoint and understanding, that the libraries cannot functionally exist without an engaged and qualified staff, who are connected to one of the local libraries on a daily basis.

By Carsten Nicolaisen, Head & Hanne M. Sørensen, Deputy head of the Library Sønderborg.

vandalism is not unknown in countries outside Scandinavia. In the United States, for example, Wallin (2015) concludes about the open library in Yacolt:

> *So far, we have no indication that people are stealing stuff from us. So far, no one has broken any windows or spray-painted graffiti. We have no reason to believe the bathroom is being used as a drug den, or that teenagers are making out in corners, or that homeless people are sleeping in the lobby. The reason for this, I believe, is that the library is well-used.*
>
> *Wallin (2015, p. 13).*

These results clearly indicate that the concerns of many professional librarians were dashed. The minor problems with vandalism are often interpreted as a sign of a widespread sense of responsibility and a common feeling of ownership toward the public library in the population.

What Went Wrong: Open Libraries With Problems

Also other problems than vandalism and violence can prevent a successful implementation of an open library project. I have chosen two cases where the staff-less model apparently has not worked and where the circumstances have been publicly reported in library journals. The two cases are:

- the Ikast-Brande case (Mønsted, 2012)
- the Kolding case (Holmgaard Larsen, 2011a, 2011b)

The Ikast-Brande Case

Ikast-Brande is a municipality in Jutland in Denmark. In April 2011, it was decided by the municipal council to transform three branch libraries (Bording, Ejstrupholm, and Engesvang) into 100% staff-less branches. It was also decided that volunteers should be present in the libraries 3 h each week. In an interview (Mønsted, 2012), the library director, Martin Lundsgaard-Leth, explains that the alternative, the closing down of the three branches, had been worse, emphasizing that the decision was made by local politicians. The most interesting aspect of the case was, however, what happened afterward. Here, both loans and numbers of visits went down in all three branches. The decrease in lending was drastic: 20%. Lundsgaard-Leth interprets the decrease as the combined effect of several causes. The collection was no more attractive because it had not been renewed and kept up to date, and the receding number of visits was a part of a long-term recession. The concept of the library as a local meeting place had not been particularly successful. The disheartening experiences with the three branch libraries, however, have not created a hostile attitude toward the open library concept. On the contrary, Lundsgaard-Leth emphasizes that the concept works well in another branch library, Nørre Snede, in the municipality. Here, the staff is preserved two times a week. Together with users and volunteers, the librarian in Nørre Snede plans events and exhibitions in the library. What we can learn from the Ikast-Brande case is that a 100% or always staff-less library is a bad idea. The question whether it had been better than to close down all three branch libraries instead is, however, not as simple as it seems because an already closed down branch is, probably, often more difficult to reopen than an unmanned. But if the user base has disappeared. Indeed, the problem is more complex than that.

The Kolding Case

Kolding is a Danish provincial town with about 60,000 inhabitants in the municipality. It has been mentioned as a site where the establishment of open libraries has been problematic (Holmgaard Larsen, 2011b). The present library director, Pia Friis, tells that in 2010 three open libraries were opened in three small towns: Christiansfeld, Lunderskov, and Vamdrup. The three branch libraries were staffed twice a week about 10h per library. The Kolding municipality library system was then in the autumn of 2009 faced with demands for savings of about one million Danish crowns (about 135,000 Euros). On the other hand, about 160,000 euros were invested in establishing the three new open libraries. Unlike, in Ikast-Brande, the new staff-less libraries did not experience drastic reduction in loans and visits in 2010. On the contrary, the number of visits grew significantly. Furthermore, a national benchmark survey in 2011 revealed that the users were happy to use the branch libraries in the prolonged 80h a week. Like in Sweden, the users also very much appreciated their local branch library and the associated staff. So the open library implementation surpassed all expectations.

The librarians' union, however, at that time meant that is was irresponsible, in general, to implement open libraries. Furthermore, a few years ago, in 2007, there had been a major administrative reform in Denmark where about 300 local municipalities had been merged into 98 local, administrative units. Deep-seated resentments over "perceived discrimination" were still alive and the feeling that the former independent local communities were taken over by the larger Kolding municipality was widespread. An important element concerned the use of volunteers where the libraries then were not ready to deal with volunteers, although money had been set aside for citizens participatory activities. In fact, it is only now that the libraries are ready to work more targeted with utilizing the volunteers in the local communities. Also the cooperation with the elementary school of Christiansfeld around a common public/school library had been exemplary.

So it is a bit of a conundrum to guess what has caused the open libraries in Christiansfeld, Lunderskov, and Vamdrup to figure as particularly problematic open libraries. Indeed, in my own report (Johannsen, 2012a), the library in Vamdrup has been emphasized as a shining light and as the only indoor public space in a small town.

User Satisfaction

It has been claimed many times, especially, by supporters of the open library concept that the users are very satisfied with the model. Library director, Carsten Nicolaisen, Sønderborg Library, in 2013, said:

> *It is a daily pleasure to meet citizens who think that the open library model represent some of the best developments which has happened in the libraries for many years. I do not think, I have not had any negative reactions (from the users).*
>
> *Translated by CGJ—Jørgensen (2013).*

Behind the claim that there also seems to be solid evidence based on representative surveys. A survey conducted by the highest Danish library authority, Kulturstyrelsen, showed that 64% of the user respondents said that they visited the public library more after the introduction of the staff-less library model; further the same survey showed that 92% of the users were satisfied with the new possibilities to visit their library whenever they wanted to do so (Jørgensen, 2013). Furthermore, an extensive survey based on a representative sample of Danish users, *Fremtidens biblioteker* [*The libraries of the future*] (clearly, indicate that flexible opening hours are among the most prioritized aspects by Danish library users (see also User types and segments).

In other countries, a variety of methods and approaches have been used to measure both usage and user satisfaction. An often applied approach uses an indirect method where, for example, librarians and library directors are asked whether they feel that their users have been satisfied with the staff-less library concept. In Sweden, a 2015 survey (Johansson et al., 2015) covered the Skåne province.

User Movement Patterns

Many visitors in supermarkets have experienced that the distribution of groceries is anything but random. Indeed, commodities that you need on a daily basis like dairy products are often placed at the farthest end of the store. The purpose is, obviously, that the customer should have the opportunity to notice and become tempted to buy as many of the store's commodities as possible.

In the autumn 2011, a Danish library student, Freyja Arnardottir, conducted an interesting study at a small open library in a village library in Tune in Denmark (Johannsen, 2012a). She had obtained access to data from the library's video surveillance system where she could watch the movement patterns of 40 users (30 women and 14 men and a lot of children) in 1 week in October 2011. Based on these observations, Freyja was able to distinguish between four general movement patterns:

1. The drive-in user who quickly picks up the needed or ordered material and then after a few minutes leave the library
2. The user searching for inspiration
3. The user accompanied by children (e.g., daytime carers and families)
4. The nonloaners (who use the library for other purposes than to borrow library materials)

Based on her observations, and supplementary interviews, Freyja Arnardottir made a list or a catalog of ideas containing suggestions as to signage, improved lighting, increased use of bright colors, acquisition of a special shelf for exhibitions, new furnitures, thematic exhibitions, recommendation from the staff, and from members of the local society, events, and the creation of a special sound inside the library. Furthermore, the observations gave rise to the considerations concerning how to influence the routes around the library chosen by the users in, for example, a serendipity friendly manner.

We will now summarize the findings of this Chapter.

Summary

The overall purpose of the fourth chapter was to address the research question:

> *Which role has local communities and the users played in the development of open libraries in different countries?*

Pessimism and Optimism

The answers are typically built upon smaller or larger empirical surveys, which focus on particular problems and sometimes very specific contexts. However, across the many specific surveys certain recurrent patterns appear. For example, the widespread absence of vandalism and theft across rural and urban neighborhoods and across different nations and cultures. Another recurrent pattern has been a considerable gap between pessimistic expectations of the library professionals and the positive experiences and outcomes of implemented open library models.

Rural or Urban

In the first two sections, *urbanization* and *neighborhood characteristics* we focused in particular on the traditional distinction between rural and urban environments where the conclusion was that open libraries on a global scale is not and never had been exclusively associated with rural or suburban contexts.

Men and Ethnic Minorities

Under the *sociodemographic* headline it was pointed out that the open libraries in some contexts had attracted men more than usual in public libraries. In certain urban neighborhoods, the open libraries, in particular, attracted the ethnic minorities more than the majority population. Also the finding that open library use is more related to work functions than staffed library use is interesting, although the difference is maybe more related to the time of the day than to whether the library is staffed or not.

National Puzzles

As to the *national variations* theme a number of interesting questions are still not fully answered, for example, why a rich, democracy like Germany has not adopted the open library model?

Users Are Both Fans of Staff and Open Libraries

Regarding *user types and segments*, representative surveys revealed, among others, that the most enthusiastic library user segment also seems to appreciate the open library model and perhaps the most interesting observation that appreciation of open libraries and appreciation of work of the staffs often appear together. A nearby interpretation could be that the staffed and the open library model cover different but both legitimate user needs.

A Balance Between Staffed and Unstaffed is Important

Another point based on the many empirical surveys is that until now it is difficult to find surveys that indicate that users prefer more staffed opening to 24/7 schemes. On the other hand, there is neither much evidence to support that totally staff-less solutions work. The idea of the desirable in a balance between staffed and unstaffed is furthermore supported by visits and loans figures.

Open Libraries Can Lead to Increased Library Use

Concerning library usage it is also an interesting finding that open libraries can often lead to increased library usage especially among busy commuters.

Open Libraries Are Supported by Representative Surveys

As the last point, it should be mentioned that the benefits of open libraries increasingly find support in representative, national surveys. And as the very last point: open libraries are in many countries, for example, the United Kingdom, Sweden, and Denmark, after thorough user surveys and user consultations and not as isolated top-down procedures.

Critical Success Factors

5

Introduction

This chapter will begin with a presentation of a number of strategies, illustrated by practical library examples and experiences, to optimize the success of open libraries (Johannsen 2012a, 2012b). Such strategies are aimed at affecting the "critical success factors." The term "best practices" is more action-oriented, focusing on what different libraries had done to meet the challenges, set by the critical success factors. Finally, the concept "lessons learned" represents the wisdom of failures. Indeed, bad experiences should not be forgotten as fast as possible. Sometimes, it turns out that what represents a failure in one context can show up to be a successful feature in another. The underlying causes of failures can also often be interpreted differently.

Let us consider a practical example. During the first years of open libraries in Denmark, the type was considered to fit rural and suburban environments, but it would apparently not work in cities and urban contexts. Although the open library concept had already been successfully applied to densely populated districts in Taiwan, it was thought that such experiences were not relevant in a Scandinavian context. Later, despite the initial skepticism, staff-less libraries were introduced in Danish urban areas and it turned out that a few libraries experienced vandalism and unrest during the unstaffed opening hours. The involved libraries, however, tried in different ways to counteract the problems, mostly with success. In very few cases, the open libraries were closed and only staffed opening hours were allowed.

The point is that this simple case, apparently, contains many different examples of best practices and lessons learned. The most relevant critical success factor is certainly that the character of the environment and the neighborhood of the library are critical success factors. An example of a best practice could be the successful strategies chosen by the involved library to lower the temperature in urban neighborhoods through, for example, changed opening hours, whereas a lesson learnt could be never to set up an open library in certain urban areas. Compared to the best practice idea, the lessons learnt concept represent a more negative and pessimistic attitude. We have already tried that, and it did not work, so: no thanks. If the recommended best practice does not work, experiments with other methods are carried out.

The strategies can be divided into four broad categories:

1. Strategies to prevent vandalism, unrest, and harassment
2. Strategies to make the library an inspiring and inviting place to stay
3. Strategies to improve user friendliness
4. Strategies to promote the library as a local third place or meeting place (Johannsen 2012a, 2012b, p. 339)

Staff-Less Libraries. http://dx.doi.org/10.1016/B978-0-08-101923-8.00005-8

Strategies to Prevent Vandalism, Unrest, and Harassment

When the idea of open libraries was introduced to the Danish library profession around 2009–10, many skeptical voices were heard, fearing widespread vandalism and destruction of library materials, furniture, and buildings. It turned out, however, that the very pessimistic expectations were mostly wrong. Very few examples of vandalism were actually seen. A widespread explanation was the lack of vandalism and harassment because the staff-less libraries were only placed in peaceful surroundings such as small villages and suburban neighborhoods. Although cases of vandalism occurred when staff-less libraries were introduced in cities and urban areas, they primarily occurred in Copenhagen and in Aalborg.

Vandalism and unrest in open libraries can act in many different ways and also have different causes, for example, boredom, poverty, ethnic tensions, gang crime, and the like.

Counterstrategies include

- Changing of opening hours
- Rules and restrictions of access to certain age groups
- Cooperation with local police, social workers and authorities
- Surveillance technology and monitoring
- Information to the public about surveillance technology and monitoring

Strategies to Make the Library an Inviting and Inspiring Place

Many different kinds of people visit a public library: children, parents, kindergarten teachers, grandparents, students, single women, men, etc. Some of the groups and individuals have quite different needs and requirements as to security and trust. The same is true for what makes a place an inspiring and inviting place. Nevertheless, there is a number of means that most people think will make the open library appear as an inviting and inspiring place. Enough light, for example, is essential, especially through the dark hours of the winter. But also the sound scape can play an important role. In the libraries in Aalborg, for example, experiments with background music in open libraries have taken place (Holmgaard Larsen, 2013). Of course, libraries in designing their open library sound-scapes should take care not to be confused with elevator music or muzak of supermarkets and shopping malls.

Besides the diversity of the staff-less library visitors, it is also essential to consider the increased importance of the library's interior design and, very important, that it had been cleaned up recently. In a staffed library it often happens that the effect of a less lucky interior design or a messy disorder is counteracted by a friendly and service-minded staff. In the staff-less library, however, the user's impression of the physical facilities and of the security of the place is paramount.

Nudging means that you instead of strict rules and regulations try to affect the behavior of the user indirectly. Nudging is an obvious technique to use in an open library.

The Dark Entrance—The Jægersborg Case

A public library example (Johannsen, 2012a) may illustrate the point. In the northern part of Copenhagen, there is a public library, Jægersborg library, which in 2011, was transformed into the first open library of the municipality. The building of the library was from 1963 and designed by a recognized library architect, Flemming Lassen (1902–84) (Dahlkild, 2011). The interesting feature in this context is, however, the entrance area, a beautiful forecourt, planted with Wisteria and Emperor Bush. The problem was, however, that this forecourt was not illuminated. To especially elderly users, passing this dark forecourt during the staff-less opening hours appeared to be a scary experience. The case was, however, hampered by restrictions as to changing the work of the famous architect. Eventually, with the help of another architect an elegant solution was found, using a square light fixture, making the entrance light, and inviting without compromising the original architectural qualities. And the Jægersborg open library has ever since been a very popular place to stay for all both during the staffed and unstaffed opening hours. The Jægersborg case illustrates how small details regarding the distribution of light and darkness can play a major role, also when the dark matter is found outside the library.

A similar observation concerning, the importance of apparently small details, points out that

> *a glass door can do wonders for the patrons feeling safe walking into a building.*
>
> *Wallin (2015, p. 14).*

There is, naturally, an ongoing professional discussion about the borders of security. Some professionals maintain that security can also create barriers. As a rule of thumb, Wallin suggests that unstaffed libraries should apply the rule, which says, that

> *until you have a reason to lock it down, leave it open.*
>
> *Wallin (2015).*

To sum up, the following elements are especially important to consider in an open library:

- Light
- Cleaning
- Clarity, from the outside, panoptical library design
- Clarity, internal, avoid claustrophobic interior design with dark corners
- Slanted shelves—colors
- Signals communicating coziness, e.g., biscuits for Christmas, coffee, and the like
- Nudging

Strategies to Improve Communication and User Friendly-ness

In 2011, the applied principles of how to make the local branch libraries fit for becoming staff-less in Denmark frequently included a simplified classification of the books and other materials accompanied by drastic reductions of the collections. Materials that had been lent less than three times within the last year could, for example, be discarded. Furthermore, all shelves were consistently and without mercy reduced to a maximum 1.6 m and all book cases reduced from five to four shelves (Johannsen, 2012a).

In sum:

- Low shelves
- No tightly packed shelves—avoid the "antiquarian bookseller" atmosphere, instead introduce effective, some would say: reckless, disposal of less used materials
- Appropriate furnishing (armchairs, working tables as needed)
- Strategic and eye-catching location of library materials, increased sensitivity toward user movement patterns
- Serendipity promoting interior design
- Varying exhibitions
- Personalized login (like in hotels: Welcome Jack Johnson)

The strategies just described have been quite general and sometimes may also be a bit overlapping and contradictory. A strategy to prevent vandalism containing strict rules and prohibitive prescripts may affect the user friendliness in a negative way and steps to facilitate the open library as a meeting place may bother the individual user. So just as during the staffed hours the management of unstaffed opening hours contains its own challenges, paradoxes, and problems. Thus the above-mentioned strategies are more thought of as ideas and suggestions than as strict rules to follow.

Strategies to Promote the Library as a Local Third Place or Meeting Place

The promotion of the staff-less library as a local meeting place concerns both the interior design and also a cooperative attitude toward local associations and cultural activities and actors. Networking, marketing, and, of course, adequate meeting facilities are all central factors.

Summary

The chapter has dealt with the research question: "which critical succes factors and associate strategies have until now been developed?"

In chapter 5, four generic strategies were defined: Strategies to prevent vandalism, unrest, and harassment; strategies to make the library an inspiring and inviting place to stay; strategies to improve user friendliness; and strategies to promote the library as a local third place or meeting place. Besides, examples and cases of how the different strategies could look were presented based on practical experiences.

To counteract vandalism, libraries could for example

- Change the opening hours
- Apply rules and restrictions of access to certain age groups
- Provide cooperation with local police, social workers, and authorities
- Provide surveillance technology and monitoring
- Provide information to the public about surveillance technology and monitoring

To make the library an inspiring and inviting place to stay the libraries could focus on

- Lighting
- Cleaning
- Clarity, from the outside, panoptical library design
- Clarity, internal, avoidance of claustrophobic interior design with dark corners
- Slanted shelves, colors
- Signals communicating coziness, e.g., biscuits for Christmas, coffee, and the like
- Nudging

To improve user friendliness a strategy could include

- Low shelves
- No tightly packed shelves, avoidance of an antiquarian bookseller atmosphere, effective, some would say: reckless, disposal of less used materials
- Appropriate furnishing (armchairs, working tables as needed)
- Strategic and eye-catching location of library materials, increased sensitivity toward user movement patterns
- Serendipity promoting interior design
- Varying exhibitions
- Personalized login (such as in hotels: Welcome Jack Johnson)

Finally, to promote the library as a local third place or meeting place, the library, for example, can

- Network
- Express a cooperative attitude toward local associations and cultural activities
- Market its offerings
- Provide adequate and up-to-date meeting facilities (interior design)

We will now continue with the topic on implementation from beginning to the completed open library in the next chapter.

A Step-by-Step Approach to Implementation

6

Introduction

We will now go through the steps leading to an implementation of an open library solution in a public library. A significant input to the chapter comes from companies within the field, Bibliotheca and Cordura. Both have kindly made available valuable descriptions, analyses, and caveats based on their long and comprehensive international experience within the field. Also, a number of examples will illustrate how the specific processes have elapsed in different libraries. The implementation process can be divided into consecutive phases each with different focus, purposes, goals, and associated milestones.

The first phase, typically, contains the selection and planning of the technological infrastructure. Then or at the same time may follow an important phase of internal marketing. Then follows the design and furnishing of the library. During the first two or three steps a well-functioning and flexible cooperation with an experienced software and hardware provider such as, for example, Axiell, Bibliotheca, and Cordura, which have assisted open library building worldwide, is especially essential. The following description is very much based on inputs from Bibliotheca and Cordura complemented with cases either provided directly from the involved libraries or through Bibliotheca and Cordura.

Selecting and Planning the Technological Infrastructure

The phase includes a number of consecutive steps and processes.

Order Process

The official process begins, typically, with the dialog between the customer and the supplier. Suppliers, here, offer a consultancy service where they make sure that the customer is getting the best solution with the needed components for the individual library. Very often this is initiated with a site survey of the library already before the offer is provided.

Start-Up Meeting

The project begins with a start-up meeting, where all participants attend. It is at this meeting that the open library solution is presented to external participants, the prior

Staff-Less Libraries. http://dx.doi.org/10.1016/B978-0-08-101923-8.00006-X

installations are clarified, integrations are discussed, and interfaces are agreed upon. The further stages of the project and the time schedule are also determined at this meeting.

The open library supplier is normally also the project manager of the open library project. But it is always recommended that the library also appoints a project manager from the library for the project. In addition, the project participants normally consist of an electrician regarding the cabling, a representative from the alarm company regarding the alarm integration, a representative from the door supplier regarding the connection of the outer door to the solution, and a representative from the library's information technology department regarding the integration to the library management system, Internet Protocol addresses for the equipment, etc.

After the first meeting the project manager provides a summary from the meeting. All participants should give their approval and the project will then proceed according to the plan.

Preparation and Installation

All participants in the project have different tasks in relation to the project. Common to all the tasks is that they must be ready before the installation of the open library begins. On the day of the installation all participants should be present so the final tests of the solution can be done and approved by the library.

Apart from the installations concerning the physical installation of the open library, the library also has to take account of changes in the physical library in relation to the new use of the library, without staff present; for example, ensuring that the equipment, such as telephones, staff computers, and other equipment on the desk, works. Sometimes certain offices are locked during the staff-less opening hours. They should also consider the decor of the room, so that it is usable without the staff present.

It is the impression of the suppliers that the libraries, typically, take this task very seriously because they care for their users and their library.

In Operation

After the installation most commonly there will be a trial period, where the library can try and test the solution and make the last preparations for the inauguration of their new Open Library. In this period, the library will also gain access to the administration program for the Open Library. From this program, the library can control and edit almost any setting in their Open Library.

Wishes and Requirements for the Open Library

There is a big difference between the wishes, requirements, and rules for the open library between the different libraries and within the different countries, in particular, concerning the requirements for monitoring (such as when the cameras should be recording or is monitoring even required) and the entry requirements (such as age restriction and rules for validation). As mentioned, access to video surveillance is much more restricted in Sweden than in Denmark. Some of the differences are due to the different laws in the countries, some has to do with the location of the library, some reflects the libraries' view of the users and the local community and so forth.

The chapters mentioned earlier have, hopefully, provided arguments to convince both staff and local politicians of the benefits to be achieved through the implementation of an open library solution in a particular municipality.

Length of the Implementation Process

The experiences about how long the implementation process will last can vary a lot both according to local decision-making traditions and procedures and because decision-making processes, especially, in the public sector can be particularly muddy. Sometimes decision-making processes in the public sector are anything but rational. According to the authors of the famous "garbage can" model, decision-making processes in public and educational organizations are often characterized as "a collection of choices looking for problems" and "solution looking for issues to which they might be an answer" than rational calculations of means to achieve a specific end or goal (Cohen, March, & Olsen, 1972, p. 1). Also open libraries can be seen as solutions on many different problems: commuters with needs for extended library opening hours and local communities threatened by depopulation and maybe also here and there financial problems and savings. Nevertheless, we have decided to look at the problems in a natural order starting with selecting and planning a technological infrastructure.

The main tasks of the technology vary somewhat depending on the both local tradition and networks and the legislation in the country concerned. In Sweden, for example, it is much more difficult to install closed-circuit television (CCTV) and video surveillance in a library than in Denmark because of more restrictive laws in this regard in Sweden. The typical tasks include:

- Design and installation of an entry and access system
- Design and installation of CCTV and video monitors
- Defining a policy as to how long video recordings are kept
- Decisions regarding maintenance and repairs
- Choice and agreements with hard- and software suppliers

A planner mentions, for example, that a special challenge in the automation process occurs when the doors are very different—from old wooden arched doors to glass doors. Some recommends to conduct pilot tests in two completely different buildings.

Internal Marketing

As has been mentioned already a certain skepticism toward the short- and long-term effects of an open library implementation can create worries and resistance among staff members. According to experiences from more countries this resistance will often disappear after the implementation. However, it is important that the leadership of the open library implementing library takes the staff concerns serious and starts, for example, adequate information campaigns, library visits, and the like.

Designing and Furnishing the Library

Designing and furnishing is also an important challenge. Often in a municipality with several libraries this task is partly centralized through the establishment of a special task force. It is often necessary to establish such a brutal task force with authority to make decisions because some of the open library principles may run counter to basic assumption within a traditional library culture. Extensive disposal of books and other library materials is often necessary to create overview and lower shelving. Often there is also a need for a more visible exposure of certain titles in an open library during the staff-less opening ours. The presentation of the title pages of books instead of the usual backs of the books can inspire but is also much more bulky.

One thing is technical functionality, another architectural-aesthetic quality. As design more and more becomes an important element of an organizations identity and image, a need emerges to integrate the design of a library's self-service infrastructure into a harmonious whole including automates, info-screens, shelves, and the like. Some software companies like, for example, Axiell, here, offer to help libraries to integrate their self-service equipment into an aesthetic valid solution. In Denmark, the company in the summer 2014 assisted the research library Danmarks Kunstbibliotek (Denmark's Art Library) to create an aesthetic appealing solution in cooperation with an architecture firm (http://axiell-danmark-as).

Often when a local library changes its status from a staffed library with relatively few opening hours to an open library with many, different local actors may find the library more interesting and attractive and the new situation invites and opens up for new cooperation models.

Optimizing Serendipity and Variety (e.g., Exhibitions)

When redesigning an "ordinary" library into an open library a number of different smaller or bigger changes can be considered. Some of them have something to do with enhancing the security whereas others are focused on optimizing the serendipity within the library. Serendipity concerns the opportunity to make unexpected but nevertheless useful finds. Typically, serendipity occurs when you find an author or a book title that you barely did not know existed but which you now almost cannot do without. Serendipity can certainly occur both in staffed and in unstaffed library environments. In unstaffed contexts, however, the challenges are even bigger because the applied means are more restricted.

Typically, small exhibitions and shelves exposing the title pages of the books are used to stimulate serendipity in open libraries. Many librarians emphasize the importance of continually changing and updating exhibitions and the exposed title pages.

Another often applied serendipity optimizing strategy could be to have posters where named staff members or users recommend certain book titles, authors, genres, and the like.

Today, however, new media have created quite new technical opportunities to achieve both more personalized and entertaining environments in the open library during the staff-less opening hours. intelligent screens that can sense the presence of a particular user and show that personalized messages or suggestions are among the many new opportunities.

Inviting and Facilitating Local Support to Use the Library as Meeting Point and Third Place

Here networking certainly plays a significant role. It is also important to be aware of what the library has to offer the local associations in terms of meeting facilities, opening hours, capacities, and rooms for meetings or lectures.

Summary

The chapter very much shares the character of a check list, remember this, be aware of that. It could with advantage be read together with the section on technological requirements in Chapter 3. As one-size-fits-all solution are very rare within the open library business, it would have been both difficult and maybe also a bit problematic with a universal to-do-list.

Conclusions

7

In this final chapter, I will sum up the considerations made in the book and also provide them with some wider perspectives.

The first research question concerned the characteristics of the benefits and the challenges of the open public libraries and about the particular aspects of the open libraries that are controversial in professional discourses.

The benefits or the pros of the open libraries are primarily two. The open libraries make the services of the public libraries available for more people. They radically increase the possibilities for visiting a library when the user wants it or when the user is free to do it. Especially commuters and people with odd or unusual work duties will benefit from the open libraries.

Second, the open libraries can contribute to maintaining threatened local communities by providing them with, among other things, social capital. The primary objections against open libraries point out that the staff-less libraries lack the availability of assistance from staff when visiting a library. In most cases, however, libraries are not staff-less all the time. So if you need staff assistance you can use the library during the staffed openings hours. Another open library counterargument claims that saving purposes lie behind open libraries and that makes them illegitimate. This is not a particularly strong argument as long as a proportion of staffed opening hours are maintained.

Chapter 2 focused on the two questions:

- "How has the staff-less library concept developed and how is this development to be seen in a broader library history context?"
- "How has the open library concept developed and spread in seven different countries in terms of primarily quantitative measures?"

As to the first question, my contribution is that I view the open library development as a public library revolution complementing and depending on two earlier library service revolutions: the open access and the automated self-delivery and self-checkout implementation of the 1990s.

Regarding the second question it is emphasized that the open library concept emerged for the first time in two opposite corners of the Earth: Denmark and the Far East around 2004. Today, the open libraries are found primarily in the Nordic countries and the United Kingdom and in the Far East.

Chapter 3 focused on the three questions:

- "What characterizes the development of open library technological infra-structure?"
- "What characterizes the themes and findings of reports and research about open libraries in five different countries?"
- What characterizes the different reactions of the principal public library stakeholders (e.g., local politicians, users, professional staff) upon the open library development?

Staff-Less Libraries. http://dx.doi.org/10.1016/B978-0-08-101923-8.00007-1

As to the open library technology, it was emphasized that the open library concept relies on advanced technological infrastructures and that a proper solution requires the integration of several different systems. Particularly important is the radio frequency identification technology.

As to the different reports on open libraries from the Nordic countries and the United Kingdom a lot of interesting problems have been raised and many research results achieved. A particularly interesting finding is the realization that users do not appear to see any contradiction between their appreciation of the services of the library staffs and their excitement of the open library model. Many users are not ready to lose the staff to totally get an open library solution. On the other hand, there is no evidence to support that they in general prefer more staffed opening hours rather than an open library solution.

Concerning the third question on stakeholders, it was concluded that a certain pattern had shown up in different countries with a winning pro-open library coalition consisting of the users, the local politicians, the local press, and library leaders and many staff members in opposition to some librarians and trade unions with national library organization as neutral toward open library positive positions.

In Chapter 4 only one question is addressed:

- Which role has the local communities and the users played in the development of open libraries in different countries? (Chapter 4)

The chapter relied much on result from different user studies prepared by the author and by others. Many different findings were presented—some surprising, others less surprising. One of the big issues concerns whether open libraries provide the library with new user groups or not. It appears as if men and members of ethnic minorities among other appreciate the open libraries. Another interesting observation concerns what different user segments think about the open libraries. Here, a Danish user segment called the "cultural super user" (well-educated, middle aged women) appears to be particularly enthusiastic about the open library concept. The attitude of this group is interesting in a cultural policy perspective because the group is among the most loyal and eager library users.

Chapters 5 and 6 are more practical by nature. Chapter 5 deals with critical success factors (CSF) and associate strategies that have until now been developed. Chapter 6 deals with the guidelines that are useful when implementing an open library services. As to the CSF four are identified and associated strategies have been suggested. The four strategies are

1. Strategies to prevent vandalism, unrest, and harassment
2. Strategies to make the library an inspiring and inviting place to stay
3. Strategies to improve user friendliness
4. Strategies to promote the library as a local third place or meeting place (Johannsen, 2012a, 2012b, p. 339)

References

Aabø, S., & Audunson, R. (2012). Use of library space as and the library as place. *Library & Information Science Research*, *34*, 138–149.

Anonymous. (2004). Free library of Philadelphia, U.N.C. – Greensboro choose Sirsi. *Advanced Technology Libraries*, *33*(8), 5–7.

Anonymous. (2008). 3M library systems: RFID Technik für Bibliotheken in Paris. *BuB: Forum Bibliothek und Information*, *60*(6), 407.

Anonymous. (2009a). San Diego public libraries rolls out self-service machines. *Library Hi Tech News*, *26*(3/4), 29.

Anonymous. (2009b). San Francisco quickens return turnaround. *American Libraries*, *39*(6), 113.

Anonymous. (2009c). Sunnyvale library upgrades to Bibliotheca – RFID solutions. *Managing Information*, *16*(5), 39.

Anonymous. (2009d). Sicherung für konsolenspiel. *BuB: Forum Bibliothek und Information*, *61*(3), 181.

Anonymous. (2011a). News from the world: Croatia. *Focus on International Library & Information Work*, *42*(1), 29–31.

Anonymous. (2011b). Norway. *Scandinavian Public library Quarterly*, *44*(1), 26.

Anonymous. (2012a). Global reach. *American Libraries*, *43*(9/10), 13–14.

Anonymous. (2012b). Ottawa public libraries chooses Bibliotheca. *Advanced Technology Libraries*, *41*(6), 9–11.

Audunson, R. (2005). The public library as a meeting place in a multicultural and digital context: the necessity of low-intensive meeting-places. *Journal of Documentation*, *61*(3), 429–441.

Bacchi, C. (2009). *Analysing policy: What's the problem represented to be?* French Forest, NSW: Pearson.

Baruchson-Arbib, S. (1997). The 'self-help section' in public libraries: the case of Israel. *Public Library Quarterly*, *16*(3), 41–49. Retrieved from http://search.proquest.com/docview/57435074?accountid=13607.

Biblioteket Sønderborg. (2016). *Udkast til Biblioteksplan 2015–2020 [Draft on library plan 2015–2020].*

Björneborn, L. (2008). Serendipity dimensions and users information behaviour in the physical library interface. *Information Research*, *13*(4), 38.

Black, A. (1994). The open access revolution in British libraries: consumer democracy or controlling discourse. *Librarians World*, *5*, 3–8.

Blomqvist, J. (2015). *Henkilökunnan kokemuksia omatoimikirjastoista (The staff's experiences of the self-service libraries).* Turku University of Applied Sciences. https://publications.theseus.fi/handle/10024/101715.

Broe Sørensen, L. (2013). *Bedre biblioteker 2013 – Nationale resultater [Better libraries 2013 – National results.* Rambøll.

Brukeradfærd på storbybibliotekerne [User behavior at the city libraries] (2015). Norway.

Cohen, M. D., March, J. G., & Olsen, J. P. (1972). A garbage can model of organizational choice. *Administrative Science Quarterly*, *17*(1), 1–25.

Cookman, N., & Burgess, H. (1996). 3M self issue terminal at Maidenhead library. *Vine*, *105*, 20–23. Retrieved from http://search.proquest.com/docview/57433696?accountid=13607.

Dahlkild, N. (2011). *Biblioteket i tid og rum: Arkitektur, indretning og formidling [The Library in Time and Space: Architecture, Design and Communication]*. Copenhagen: Danmarks Biblioteksforening.

Dahlkild, N. (2015). Kulturens huse: Arkiver, biblioteker, museer og kulturcentre [The houses of culture: archives, libraries, museums and culture centres]. In N. Dahlkild (Ed.), *Huse der har formet os: Arkitekturhistorien bag danskernes institutioner og offentlige rum [Houses that have influenced us: The history of architecture behind Danish institutions and public spaces]* (pp. 61–96). Copenhagen: Museum Tusculanums Forlag, Danmarks Kunstbibliotek.

Deildok, M. (2010). Bibliotekaren marts.

Ditmansen, H., Egeland, L. W., & Sama, T. (2014). Når vertinnen er gått hjem [When the hostess has left]. *Bibliotheca Nova*, *3*, 80–89.

Drost, P. (May 8, 2013). Primitive spørgsmål [Primitive questions]. *Perspektiv*.

Eggink, G. (2000). Bibliotheek Amsterdam schaft dertig lendomaten aan. [Amsterdam public library instals 30 self-service lending machines]. *BibliotheekBlad*, *4*(1), 10–11. Retrieved from http://search.proquest.com/docview/57496207?accountid=13607.

Eklof, A. (2013). Understanding information anxiety and how academic librarians can minimize its effects. *Public Services Quarterly*, *9*(3), 246–258. http://dx.doi.org/10.1080/15228959.2013.815529.

Elding, K., & Krell, A. (2015). Meröppna bibliotek – utveckling, inte avveckling [Staff-less library – development, not dismantling]. *Biblioteksbladet*, *9*, 12.

Engström, L. (2015). Bara meröppet? [Just staff-less?]. *Biblioteksbladet*, *6–7*, 39–41.

Engström, L., & Eckerdal, J. R. (2016). In-between strengthened accessibility and economic demands – analyzing self-service libraries from a user perspective. *Journal of Documentation*, *73*(1), 145–159.

Enis, M. (2014). Queens tests job app kiosks. *Library Journal*, *139*(9), 13–14.

Florida, R. (2005). *Cities and the creative class*. New York: Routledge.

Foucault, M. (1980). *Power/knowledge : Selected interviews and other writings 1972–1977*. Brighton: Harvester P.

Goffman, E. (1990). *The presentation of self in everyday life*. London: Penguin Books.

Haase Svendsen, G. L. (2006). Studying social capital in situ: an anthropological approach. *Theory and Society*, *35*(1), 39–70.

Haase Svendsen, G. L. (2013). Public libraries as breeding grounds for bonding, bridging and Institutional Social Capital: the case of Branch Libraries in Rural Denmark. *Sociologia Ruralis*, *53*(1), 52–73.

Hall-Ellis, S. (2015). Nudges and decision making: a winning combination. *The Bottom Line*, *28*(4), 133–136. Retrieved from http://search.proquest.com/docview/1753391027?accountid=13607.

Hampshire, M., & Sanford, C. E. (2009). Library-a-go-go: bringing the library to the people. *New Library World*, *110*(11/12), 541–549.

Heuff, I. (1999). Doe-het-zelf-uitleenbalie: Bezuiniging of idealisme?: Ervaringen in Zeeland en Overijssel duiden nog niet op groot succes. [A self-service loans counter - budget saving or mere daydream?: Experiences in Zeeland and Overijssel suggest that problems have not yet been ironed out]. *BibliotheekBlad*, *3*(15), 20–22. Retrieved from http://search.proquest.com/docview/57473204?accountid=13607.

Hill, N. M. (2010). Is there room for customer service in a self-service environment? *Public Libraries, 49*(6), 13–19.

Hitchcock, G. (2016). *How technology can save council libraries*. Publiclibrariesnews.com.

Hjalmarson, P. (2015a). Meröppet ett led i nedskärningarna av folkbiblioteken [Staff-less a part of downsizing the public library]. *Biblioteksbladet, 8*, 10.

Hjalmarson, P. (2015b). Merôppet nytt concept för nedskärningar [Staff-less – a new word for downsizing]. *Biblioteksbladet, 10*, 10.

Hokka-Ahti, R. (2005). Palautuslaitteiden lumossa. [Spellbound by check-in automats]. *Kirjastolehti* (2), 13–15. Retrieved from http://search.proquest.com/docview/57607435?accountid=13607.

Holmgaard Larsen, J. (2011a). Self-service at unusual times; the "Open libraries" are popular in Denmark/Open 70 to 80 hours. *BuB Forum Bibliothek und Information, 3*, 118–119.

Holmgaard Larsen, J. (2011b). Åbne biblioteker – en dansk succeshistorie [Open libraries – a Danish success]. *Bibliotek og Medier, 2*, 14.

Holmgaard Larsen, J. (2013). Open libraries in Denmark. *Scandinavian Public Library Quarterly, 46*(3), 22–23. Retrieved from http://search.proquest.com/docview/1496968846?accountid=13607.

Hvenegaard Rasmussen, C., Jochumsen, H., & Skot-Hansen, D. (2011). *Biblioteket i byudviklingen –oplevelse, kreativitet og innovation [The library in the urban development – experience, creativity and innovation]*. Copenhagen: Danmarks Biblioteksforening.

Jakobsen, V. G. (1996). The self-service system in Toensberg public library, Norway. *Vine, 105*, 33–35. Retrieved from http://search.proquest.com/docview/57436404?accountid=13607.

Jochumsen, H., & Hvenegaard Rasmussen, C. (2000). *Gør biblioteket en forskel? [Does the library make a difference?]*. Copenhagen: Danmarks Biblioteksforening.

Johannsen, C. G. (2012a). *Åbne biblioteker – et kortlægningsprojekt [Open libraries – a mapping project]*. Copenhagen: Det Informationsvidenskabelige Akademi.

Johannsen, C. G. (2012b). Staffless libraries – recent Danish public library experiences. *New Library World, 113*(7/89), 333–342.

Johannsen, C. G. (2014a). Innovative public library services – staff-less or staff-intensive? *Library Management*, 35 (6/7), 469–480.

Johannsen, C. G. (2014b). Understanding users: from man-made typologies to computer-generated clusters. *New Library World, 115*(9/10), 412–425.

Johannsen, C. G. (2016). From clients to participants – how information technology impacts relationships between professionals and users. In D. Baker, & W. Evans (Eds.), *Chandos Digital Information Review Series. Digital Information Strategies – From applications and content to libraries and people* (pp. 157–166). Waltham, MA: Chandos.

Johansson, G., Lindberg, E., & Eckerdal, J. R. (2015). *Meröppna bibliotek – en verksamhetsform för alla? [Staff-less libraries – a concept for all?]*. Lund: Lunds Universitet. http://lup.lub.lu.se/record/5035199/file/5035201.pdf.

Jokitalo, P. (2015). Denmark. *Scandinavian Public Library Quarterly, 48*(1/2), 52–53.

Jørgensen, J. (2013). Selvbetjente biblioteker boomer [Self-service libraries are growing fast]. *Momentum*. http://www.kl.dk/Momentum/momentum2013-8-2-idl29736.

Kann-Rasmussen, N., & Tank, E. (2016). *Strategi som legitimitetsarbejde – Strategiske svar å bibliotekernes udfordringer [Strategy as legitimation work – strategic responses to the library challenges of the libraries], Nordisk Tidsskrift for Informations videnskab og Kulturformidling, 3*(5), 5–19.

Kelly, M. (2016). *Review of the open library concept in UK public libraries: Report for Bibliotheca*. Manchester: APSE (consultancy report – confidential).

Kjekstad, T. (1996). Selvbetjent utlan. [Self service lending]. *Synopsis*, *27*(2), 75–77. Retrieved from http://search.proquest.com/docview/57401171?accountid=13607.

Kraljic, M., & Maver, J. (2006). Uvajanje samopostrezne izposoje in odziv uporabnikov. [Introduction of self issue services and respond of the library users]. *Knjiznica*, *50*(3), 85–104. Retrieved from http://search.proquest.com/docview/57602323?accountid=13607.

Larsen, M. (2007). Use the library – outside opening hours as well. *Scandinavian Public Library Quarterly*, *40*(2), 6–8. Retrieved from http://search.proquest.com/docview/57613465?accountid=13607.

Lauridsen, J. (2011). Pivåbne biblioteker. *Perspektiv* Okt 29.

Lawrence, P., & Weber, L. (2012). Midnight-2.00 a.m.: what goes on at the library? *New Library World*, *113*(11/12), 528–548.

Lek Choh, N. (2003). A totally do-it-yourself library without a library customer service desk: the Singapore experience. In *IFLA conference proceedings World Library and Information Congress: 69th IFLA General Conference and Council Berlin, Germany. August 1–9, 2003* (pp. 1–6).

Lerche, A. (2010). Shall a call center replace the librarian? *Bibliotekspressen*, 9 (June 3), 24–25.

Lerche, A. (2011). Hvorfor har vi ikke tænkt på det her noget før? [Why haven't we thought about it before?]. *Perspektiv*.

Lerche, A. (2012a). Åbent – det skal vi da have [Open – that's what we want]. *Perspektiv*, 12–16 Juni.

Lerche, A. (2012b). Since Islands Brygge opened up. *Perspektiv* (6), 19–20. Retrieved from http://search.proquest.com/docview/1081858568?accountid=13607.

Meuter, M. L., et al. (2000). Self-service technologies: understanding customer satisfaction. *Journal of Marketing*, *64*(3), 50–64.

Miller, J. (2014). A comparative study of public libraries in Edinburgh and Copenhagen and their potential for social capital creation. *Libri*, *64*(4), 316–326.

Mønsted, S. (2012). We don't enter a deserted restaurant. *Perspektiv* (6), 17–18. Retrieved from http://search.proquest.com/docview/1081858564?accountid=13607.

Moos-Bjerre, M., & Moos-Bjerre, T. (2014). *Fremtidens Biblioteker – målgruppebaseret viden til biblioteksudvikling [The Libraries of the future – target group oriented knowledge for library development]*. Copenhagen: Tænketanken Fremtidens Biblioteker.

Mountain view implements innovative's self-service products. (2006). *Advanced Technology Libraries*, *35*(12), 8. Retrieved from http://search.proquest.com/docview/57680965?accountid=13607.

Mustikkamäki, M. (2015). *Undersökning om meröppet på de allmänna biblioteken i Finland*. Regional State Administrative Agencies (Both in Finnish and Swedish). https://www.avi.fi/documents/10191/3369535/Unders%C3%B6kning+om+mer%C3%B6ppet+p%C3%A5%20de+allm%C3%A4nna+biblioteken+i+Finland+2015.pdf/232e73e4-ef5c-4525-b9c0-6dceebfe8320.

Nyeng, P. (1998). Det store biblioteksspring pa Frederiksberg. [The great library leap in Frederiksberg]. *Bibliotekspressen* (6), 182–183. Retrieved from http://search.proquest.com/docview/57438925?accountid=13607.

OECD. (2015). *OECD skills studies: Data policy reviews of adult skills: Finland*. OECD. http://minedu.fi/export/sites/default/OPM/Koulutus/artikkelit/piaac/liitteet/adultskillsfinland.pdf.

Öffentlicher Bücherschrank [Public book-shelfs] (2016). Wikipedia.

Oldenburg, R. (1989). *The great good place: Cafés, coffe shops, community centers, beauty parlors, general stores, bars, hangouts and how they get you through the day*. New York: Paragon house.

Ollikainen, S. (2015). *Kohti omatoimisempaa kirjastoa : Itsepalvelun kehitys suomalaisissa kirjastoissa (English abstract) [Towards more self-directed library: The development of the self-service in Finnish libraries]*. Turku University of Applied Sciences. https://publications.theseus.fi/handle/10024/97368.

Palmer, M. (2006). Using RFID to transform Essex libraries. *Library Hi Tech News*, *23*(10), 21–22. http://dx.doi.org/10.1108/07419050610730070.

Petersson, M. (1996). TOR-IN and TOR-OUT – a Swedish automated system for issuing and returning library books. *Vine* (105), 30–32. Retrieved from http://search.proquest.com/docview/57438340?accountid=13607.

Planstyrelsen (198?). Tillægsfunktioner i landsbybutikker – et forsøg med kombinerede serviceenheder [Additional functions in village shops – a project on combined service units]. Copenhagen: Institut for Center-Planlægning.

Pohl, M., & Schubert, E. (2006). Wenn - dann 100 prozent. selbstverbuchung mit RFID in der Münchner Stadtbibliothek. If – then 100 per cent [Self-service borrowing with RFID at Munich municipal library]. *BuB Forum Bibliothek und Information*, *58*(1), 53–55. Retrieved from http://search.proquest.com/docview/57648024?accountid=13607.

Putnam, R. D. (1995). Bowling alone: America's declining social capital. *Journal of Democracy*, *6*(1), 65–78.

Putnam, R. D. (1996). The strange disappearance of Civic America. *The American Prospect*, *7/24*, 34–48.

Riess-Taggart, J. (2006). Palm Beach county library system implements intelligent media manager. *Public Libraries*, *45*(3), 17–19.

Rogers, M. (2005). GA, MD take RFID plunge. *Library Journal*, *130*(15), 26.

Sigwald, R. (2016). Self-service customer service models in libraries. *Journal of Library Administration*, *56*(4), 453–478. http://dx.doi.org/10.1080/01930826.2016.1157429.

Skåne, K. (2011). *Tack vare det meröppna biblioteket finns det ett bibliotek även för mig – en utvärdering av Veberöds meröppna bibliotek 2010–2011.*

Skot-Hansen, D., Hvenegaard Rasmussen, C., & Jochumsen, H. (2013). The role of public libraries in culture-led urban regeneration. *New Library World*, *114*(1/2), 7–19.

Slots- og kulturstyrelsen. (2015). *Folkebiblioteker i tal [Public libraries in numbers] 2014*. http://slks.dk.

Slots- og kulturstyrelsen. (2016). *Oversigt over åbne biblioteker pr. 1. September 2016 [Overview over open libraries September 1, 2016]*. slks.dk.

Smith, P. S. (2008). Self-check: a lesson in mistaken assumptions. *Computers in Libraries*, *28*(2), 14–18. Retrieved from http://search.proquest.com/docview/57621324?accountid=13607.

Steffensen, R. I. (1992). *Det brugerbetjente udlånssystem: Australsk teknologiinspiration. [The self service issue system: Australian technological inspiration]* (19. B70 (Bibliotek 70), 596–597. Retrieved from http://search.proquest.com/docview/57315447?accountid=13607.

Suchman, M. C. (1995). Managing legitimacy: strategic and institutional approaches. *The Academy of Management Review*, *20*(3), 571–610.

Tapdrup Mortensen, M. (October 28, 2011). *Den lille amerikanske flirt [The small American flirt]*. Weekendavisen.

Thaler, R. H., & Sunstein, C. R. (2009). *Nudge: Improving decisions about health, wealth, and happiness*. NewYork, NY: Penguin Books.

The Economist. (2014). *Nudge nudge, think think: The use of behavioural economics in public policy shows promise*. Available at www.economist.com/node/21551032.

Thompson, J. E. (2006). Self check-out. *Georgia Library Quarterly*, *43*(2), 15–17. Retrieved from http://search.proquest.com/docview/57646482?accountid=13607.

Thorhauge, J. (2011). Editorial: a new "open library" concept. *Scandinavian Library Quarterly*, *44*(4), 3–4.

Tseng, S., & Kuo, P. (2009). A study on the patronage of the intelligent library: The Ximen Intelligent Library of the Taipei public library system. *New Library World*, *110*(9–10), 410–429. Retrieved from http://search.proquest.com/docview/57686112?accountid=13607.

Udvalget om folkebibliotekerne i Vidensamfundet. (2010). *Folkebibliotekerne i vidensamfundet. [The public libraries in the knowledge society]*. Copenhagen: Styrelsen for Bibliotek og Medier.

Van de, B. (2002). Klanten bibliotheek Hengelo minder enthousiast over inname – en uitleenautomaten dan personeel. [Hengelo library's users less enthusiastic about the automated loans and returns machines than staff]. *BibliotheekBlad*, *6*(9), 10–12. Retrieved from http://search.proquest.com/docview/57532706?accountid=13607.

Vårheim, A. (2007). Social capital and public libraries: the need for research. *Library & Information Science Research*, *29*, 416–428.

Vårheim, A. (2009). Public libraries: places creating social capital. *Library Hi Tech*, *27*(3), 372–381.

Vårheim, A. (2011). Gracious space: library programming strategies towards immigrants as tools in the creation of social capital. *Library & Information Science Research*, *33*, 12–18.

Vårheim, A. (2014). Trust in libraries and trust in most people: social capital creation in the Public Library. *Library Quarterly*, *84*(3), 258–277.

Vårheim, A., Steinmo, S., & Eisaku, I. (2008). Do libraries matter? Public libraries and the creation of social capital. *Journal of Documentation*, *64*(6), 877–892.

Venuda, F. (2009). The new city library of Newcastle Upon Tyne, UK. *Biblioteche Oggi*, *27*(8), 47–58.

Wagman, A. K. (2011). *Olika syn på saken – Folkbiblioteket bland användare, icke-användare och personal. [Different points of view – The public library among users, non users and staff]*. Svensk Biblioteksförening.

Wallin, S. (2015). Yacolt Library Express: a model for unstaffed library service. *PNLA Quarterly*, *80*(1), 8–15.

Wieldraaijer, E. (2002a). NBD/Biblion presenteert Cor-lector als concurrent voor Tor in. the NBD/[Biblion company unveils Cor-lector in competition with Tor in]. *BibliotheekBlad*, *6*(9), 13. Retrieved fromhttp://search.proquest.com/docview/57554850?accountid=13607.

Wieldraaijer, E. (2002b). 'Geautomatiseerde uitlening en inname belangrijke instrumenten om inhoudelijke innovatie op gang te brengen'. [Automated loans and returns systems are important tools for implementing innovation]'. *BibliotheekBlad*, *6*(9), 9. Retrieved from http://search.proquest.com/docview/57528789?accountid=13607.

Wieldraaijer, E. (2005). Janice Boon Huay Tan (af)kicken op technologie. [Janice Boon Huay Tan believes in taking full advantage of technology]. *Bibliotheek* (6), 26–27. Retrieved from http://search.proquest.com/docview/57587218?accountid=13607.

Wildermann, P. (2015). *"Open library" – Der Schlüssel zum Buch [Open library – the key to the book]*. https://www.goethe.de/de/kul/bib/20676698.html.

Ytte Larsen, V. (2012). Et åbent bibliotek – et godt bibliotek? [An open library – a good library?]. *Danmarks Biblioteker*, *2*(15.3), 1.

Relevant Websites

http://www.publiclibrariesnews.com/about-public-libraries-news/unstaffed-librarie.

http://www.axiell.se/produkter-tjanster/for-lantagare/axiell-meroppet/.

http://www.helmet.fi/en-US/Libraries_and_services/Selfservice_libraries.

http://axiell-danmark-as.e-mailing.se/111928/ZgVM66dL/Abne-biblioteker-giver-tilfredse-brugere?.

http://www.fobid.nl/sites/fobid/files/FOBID%2019.16%20fin%20%20NL%20Library%20 Statistics%20%202014.pdf.

http://slks.dk/fileadmin/user_upload/0_SLKS/Fotos/Biblioteker/Biblioteker_div/8._Oversigt_ over_aabne_biblioteker.pdf.

Index

'*Note*: Page numbers followed by "f" indicate figures, "t" indicate tables, and "b" indicate boxes.'

J

K

L

M

N

O

P